ALLEGORY OLD AND NEW

ANALECTA HUSSERLIANA

THE YEARBOOK OF PHENOMENOLOGICAL RESEARCH

VOLUME XLII

Editor-in-Chief:

ANNA-TERESA TYMIENIECKA

The World Institute for Advanced Phenomenological Research and Learning
Belmont, Massachusetts

ALLEGORY VOLUMES

Book 1 Allegory Revisited: Ideals of Mankind
Book 2 Allegory Old and New: In Literature, the Fine Arts, Music and Theatre, and Its Continuity in Culture

ALLEGORY OLD AND NEW

in Literature, the Fine Arts, Music and Theatre, and Its Continuity in Culture

Edited by

MARLIES KRONEGGER

Michigan State University, East Lansing

and

ANNA-TERESA TYMIENIECKA

The World Phenomenology Institute

Published under the auspices of

The World Institute for Advanced Phenomenological Research and Learning

A-T. Tymieniecka, President

SPRINGER-SCIENCE+BUSINESS MEDIA, B.V.

Library of Congress Cataloging-in-Publication Data

Allegory old and new : in literature, fine art, music and theatre and its continuity in culture / edited by Marlies Kronegger and Anna-Teresa Tymieniecka.
p. cm. -- (Analecta Husserliana ; v. 42)
English and French.
"Published under the auspices of the World Institute for Advanced Phenomenological Research and Learning."
A selection of papers from two congresses held by the International Society of Phenomenology and Literature, the XVIIth Annual Congress in Cambridge, MA, in April, 1992, and the XXIX International Phenomenology Congress in Luxemburg, June 14, 1992.
Includes index.
ISBN 978-94-011-7649-1 ISBN 978-94-011-1946-7 (eBook)
DOI 10.1007/978-94-011-1946-7
1. Allegory--Congresses. 2. Literature--Philosophy--Congresses. 3. Arts--Congresses. I. Kronegger, Marlies, 1932- . II. Tymieniecka, Anna-Teresa. III. World Institute for Advanced Phenomenological Research and Learning. IV. International Society for Phenomenology and Literature (18th : 1992 : Cambridge, Mass.) V. International Phenomenology Congress (29th : 1992 : Luxembourg, Luxembourg) VI. Series.
B3279.H94A129 vol. 42
[PN56.A5]
142'.7 s--dc20
[809'.915] 93-8088

Printed on acid-free paper

TABLE OF CONTENTS

PART III

PART IV

ANNEX

ACKNOWLEDGEMENTS

In this volume is gathered a selection of papers from two congresses held by The International Society of Phenomenology and Literature, affiliated with The World Phenomenology Institute: the XVIIth Annual Congress in Cambridge, MA, April, 1992, and the XXIXth International Phenomenology Congress in Luxembourg, June 14, 1992.

We owe our special thanks to our Luxembourg hosts, in particular to Professor Rosemarie Kieffer, who presided at the conference, and to Professor Antoine Cipriani, President of the Luxembourg Society for Literature and Comparative Literature, who took care of the local logistics and publicity. The generous hospitality in Luxembourg was greatly appreciated by all participants.

Isabelle Houthakker merits our thanks for her copy-editing of the volume and Louis Houthakker read the proofs.

Marlies Kronegger
A-T. Tymieniecka

ALLEGORY: AESTHETIC VEHICLE OF THE SPIRIT

The turn of contemporary literature toward the concrete everyday life of average people, its search after the manifestation of their elementary instincts, feelings, and reactions before they are controlled by the mind with its moral and aesthetic molding, seems to make clear why there is now no use for apotheosis, glorification and the 'idealisation' of the 'extra-ordinary', which are characteristic of the aesthetic of allegorizing.

Concurrently, contemporary philosophy and literary theory focus upon the primary linguistic forms which make intelligible communication possible; they stay aloof from allegory, as this latter seems to be a sort of 'second degree' intellectual construct.

Allegory is, indeed, a construct of the human mind, and it is not situated, as are symbol or metaphor, at the borderline between the unintelligible and the intelligible; nonetheless its impulse stems not from the already established human mind and the intellect at work, but from the subliminal forge of the human soul on the primogenital level of the origination of the specifically human significance of life.

The actual function of allegorizing sets in when the mind has already reached the peak of its development; but its trajectory from its primogenital impulse traverses in a unifying, harmonizing fashion all the sentient, emotive, intelligible circuits of the human significance of existence.

Drawing, on the one hand, upon all three of the mind's sense-giving factors – aesthetic/poetic, moral, intellectual – it gives the decisive role to the first of them; it is essentially a manifestation of the aesthetic significance of life.

On the other hand, contrary to the tendency to see allegory as a mere construct of the intellectual faculties, it is, in fact, a *response to the deepest subliminal longings* of the human being launching upon the journey of the creative mind, longings which, concurrently with the surging of the mind, prompt the human spirit to rise and soar.

While the human spirit unfolds and takes shape in ever more elevated values, ideals, and aspirations toward generous, noble, and beautiful deeds – allowing the soul to lift its life above the concrete vicissitudes of life's struggles, and endowing life with a specifically human significance – it is from the same longing of the subliminal soul that the allegorizing function of the mind organises its work.

Thus allegory manifests itself at work when the task of the intellect, imagination, memory, will, has already been accomplished.

Indeed, allegory surges by tying the knot of the various lines of significance worked out already by the faculties of the mind while infusing them with the exaltation of the aesthetic sense; it seeks to incorporate into a cogent significant complex a selection of elevated ideals, virtues, values, which respond to the specific human longings of the period, and to project its vision into concrete life itself.

Thus an allegory allows the human soul which is deciphering it to invest life with its deepest yearnings and yet transcend life's turpitude on the wings of the spirit.

Expressing the ideals and yearnings shared by persons of the same community of life, a social group, a nation, allegory brings a given culture to its height; it manifests its highlights.

Cultures evolve, become transformed, have their peaks and their downfalls; and their respective perspectives upon the significance of life evolve. However, the human spirit which finds the apex of its manifestation in allegory will not stop rising as long as the human being remains human.

ANNA-TERESA TYMIENIECKA

PART I

The group of participants in the Allegory Conference in Luxembourg, June 1992.

MARLIES KRONEGGER

ALLEGORY AND MAXIM: POWER AND FAITH, PASSIONS AND VIRTUES

> Les grands hommes n'ont point de nation: le monde est leur patrie, toutes les coutûmes, toutes les moeurs sont naturelles & leur sient bien. Ils sont de tout païs, ils doivent savoir se transformer, quand il leur plaït, comme Protée. Ainsi, Alexandre, qui étoit né pour l'univers, ne se devoit pas moins aux peuples de l'Asie qu'à ceux de l'Europe, lui qui n'étoit au monde que pour commander. (Queen Christine, *Réflexions diverses sur la vie du grand Alexandre*)

INTRODUCTION: ALLEGORY AND MAXIM

Allegories meander like an unbroken thread through Baroque art and creativity, and likewise throughout Queen Christine's activities, festivities, and ambitions. Our goal is to present a panorama rather than a map of Queen Christine's quest as it is: a) immortalized in allegorical engravings based on historical reality and idealized in emblems and allegories of civil and religious accomplishments (these are recorded by and for the outside observer); and b) recorded in her *Maximes*, at once the personal, intimate expression and the revelation of her personal quest for transcendence and truth, a reaction to the events and turmoil of her times.

Both allegories and maxims offer a wealth of seeing, feeling, and thinking and make Queen Christine's life a fundamental paradigm of reflection on life and culture in the midst of seventeenth century Stockholm, Paris, and Rome – as observed by her contemporaries and experienced personally by Queen Christine.

As we shall see, her historical and idealized reality is illustrated in allegories of her accomplishments by contemporary artists and musicians seeking to immortalize her; and her human condition, life, and personal experience find reflection in her *Maximes*, in which she awakens and calls forth all her vital synergies in order to surpass herself, to liberate

M. Kronegger and A-T. Tymieniecka (eds.), Analecta Husserliana XLII, pp. 3–27.

herself from all constituted finite frontiers. In this study we shall focus on her quest after the ultimate destiny and significance of life: the transmutation of the empirical, aesthetic, moral, and transnatural quest for ultimate sense. In Queen Christine's own words, great men are without any specific nationality, and their quest, like hers, is to remain flexible and open to the world as was Alexander the Great. The exaltation of human qualities for her is the surpassing of all restrictions imposed by society and the revelation of the invisible, the ineffable, the transcendent reality in both art and religion. She experiences a truly spiritual ascension on the road from Stockholm, to Paris, to Rome.

I. FROM HISTORICAL REALITY TO ALLEGORICAL SUBLIMATION

Let us recall a few historical facts. The religious turmoil that permeated Europe in the sixteenth and seventeenth centuries led to the outbreak of the Thirty Years War between Catholics and Protestants in 1618. What had first seemed a war for the freedom of religion developed into a struggle by the Spanish and Austrian Hapsburgs for world supremacy, a thrust challenged and contested by France, Sweden, Denmark, and England, as they sided with the Protestant princes of Germany. The war was fought in Austria, Bohemia, Italy, the Netherlands, and Spain. The religious and national aspect was, in the end, secondary from the French perspective, as France, a great Catholic power, sided with Protestant Sweden and even called on the Turks for aid.

It was Sweden's King Gustavus Adolphus, who had saved the cause of Protestantism. As Gustavus Adolphus, victorious in the battle of Lützen, was killed in 1632, his daughter and only child, Christine, sent the regent Axel Oxenstierna to negotiate the Peace of Westphalia which ended the war in 1648. The two powers protecting the peace, France at Münster, Sweden at Osnabrük, were celebrated in a number of cultural events. The German poets Harsdörffer,[1] von Birken, and Klaj[2] and the French poet Père Le Moyne[3] celebrate in a deep emotional way, Christine of Sweden, the heroine of peace, and her father, Gustavus Adolphus, in an evocation of fireworks in their honor.[4] According to these poets, it is thanks to Christine that both peace and unity have been restored for eternity after three decades of murder.

II. THE HUMAN CONDITION: A BATTLEGROUND OF RECIPROCAL FORCES OF GOOD AND EVIL[5]

The frontispiece of the Christina Bible by D. Dirickson[6] (1646) contains an allegorical vision of Christine within the framework of historical reality – her choosing not to enter into hostilities for the arrogant pleasure of making war, but out of a noble desire to pacify Europe to bring happiness and abundance to the conquered peoples: "Ce que je désire le plus & que j'estime au-dessus de toute autre chose, c'est de pouvoir rendre la paix à la Chrétienté" (123).[7] Her crown marks the transcendent qualities of her accomplishments; its circular form indicates her perfection and participation in the divine nature, pointing to the heavens. Allegories of Religion, War, Universal Knowledge (books), and Liberty fill the four corners. The human condition is shown as a battleground of reciprocal forces of evil (the violent and chaotic) and good (religious and cultural order): Her maxim "Convertir le mal en bien est le grand art de Dieu" (XI, 17), affirms her conviction that evil has to be transformed into positive values. The approach is cool, graphic, and one might almost say abstract. "Il faut mériter une grande & belle Réputation malgré l'Envie" (II, 65) outlines her efforts, and she suggests: "Avoir l'esprit & le coeur grand, c'est avoir du mérite" (IV, 2).

The space enframed by a twisted garland, a wreath of fruits, is very clearly laid out in four horizontal parallel planes. The dark foreground is the zone of the passing army; the lighter remote background is astonishingly empty except for the sky with an angel announcing fame to the Queen, and a stork, promising good luck under the shelter of a dove to the right. The impression of the infinite extension of the landscape, the mystery of space and infinity, is enhanced by the atmospheric perspective: here Queen Christine emerges as the center of the universe, with her father, Gustavus Adolphus, as Jupiter hovering over her as an eagle. At her feet there is a lion holding a snake captive, an allegory of the ruler (the lion) conquering evil (the snake).

On the whole, the allegory is built on the twin symbols of Christian Virtues (Pity, Clemency, Truth, Generosity, Moderation) and imperial iconography (Justice, Prudence, Fortitude, Liberty). Both Fame and the Holy Spirit are hovering over the army which is moving on to further victories. Christine herself is compared to the mythical prince Hercules, the man of action, always ready to stand against the forces of evil and triumph over his enemy, which illustrates her conviction that "il y a dans la nature quelque chose d'ennemi qui s'oppose à la parfaite gloire et félicité des hommes" (VIII, 32). Queen Christine was served by two advisors, Axel Oxenstierna, representing civil power (to the left), and Lennart Torenson, the incarnation of military power (to the right) who never fought for territorial conquests (even though he had invaded Holland) but for equity, peace, and happiness.

III. CHRISTINE'S PASSIONATE ATTACHMENT TO HONOR AND VIRTUE: THE "HONNÊTE FEMME"

The French poet Le Moyne, in "*Galerie des femmes fortes*", compares her virtues to those of men on the battlefield, and indeed she is convinced that "se vaincre, c'est triompher du plus puissant de ses ennemis" (VII, 83). She had done everything to live up to her ideal of virtue. This ideal was realized in her command of various European languages, not only her native Swedish but also French, German, Italian, and Latin, which she spoke equally well, as if they were her native tongues; she read in the original editions Greek, Hebrew, and Arabic philosophy, theology, mathematics, and astronomy. One may say that her ambition was directed at making her name famous more by extraordinary merit than by political or military conquests. Her strivings were aimed more at achieving fame by her own efforts rather than through the qualities of her subjects. Hence she declared: "Le mérite & la vertu ne sont attachés, ni à l'habit, ni à la condition, ni au païs" (V, 22); "N'obéir à personne est un plus grand bonheur que celui de commander au monde entier" (V, 61); "On ne doit jamais être satisfait de soi-même quelque satisfaits qu'en paroissent les autres" (VIII, 49); and "Il faut tâcher de se surpasser toujours; cette occupation doit durer autant que la vie" (VIII, 51). These maxims outline her personal goals. Allusions to mythological characters and pictorial accounts of the glorious deeds of antiquity are evidence of the Queen's love for the scholarly achievements glorified in the allegories which endow her reign with a legendary dimension.

In 1649, J. Falk allegorically portrayed her as Minerva (Pallas Athena),[8] with a laurel-wreathed helmet, an olive branch, the symbol of Peace, and three volumes of books connoting classical learning, and the owl, signifying wisdom and scholarship. This allegorical sculpture glorifies Queen Christine's quest for intellectual perfection. As the "Minerva" of Stockholm, her strength, courage, and love of virtue have overcome the most perilous obstacles and, therefore, she has been granted immortality, the supreme recompense for human striving. According to her, "le grand secret de la vie est de se proposer un digne but & de ne le perdre jamais de vue" (II, 73); or "l'unique secret de n'être pas gouverné, c'est de croire peu & de travailler beaucoup" (III, 21); or "il ne faut compter, que sur soi-même & ses propres forces: quand elles ne suffisent pas, rien ne suffit" (VIII, 33); or "la véritable grandeur ne dépend que du coeur; quand il est grand, tout l'est" (IX, 36).

For this reason, medals and books were dedicated to the Minerva of Stockholm. When, for example, Scudéry dedicated *Alaric* to her, he surprisingly summarized her quest in an accurate way.

> La grandeur de votre âme et celle de votre esprit mettent autant Votre Majesté au-dessus des Reines, que les Reines sont au-dessus des autres Dames . . . je suis obligé d'avouer, que le Nord a maintenant sa Minerve dans Stockholm, comme il y eut autrefois une Diane dans Tauris: que l'esprit et la vertu n'ont pas de climat affecté, & qu'ils sont aussi bien à Stockholm & à Upsal que dans Rome ou dans Athènes . . . (258–260)

Her erudition explains her fondness for books and paintings. Axel Oxenstierna had inspired her lifelong effort to attain global knowledge of all of the sciences, art, and philosophy.

a) Historical events favored her intimate aspirations. The trophies won in the Swedish victory in Central Europe included the magnificent art collection of Emperor Rudolph II of its books and paintings from various European countries; the painting which she treasured most were those by the Italian masters.[9] These were sent from Prague to Stockholm. Christine asked both Dutch and French curators and librarians to classify them. Revolutions in England and France had eliminated temporarily two of her essential rivals in the international patronage stakes, Charles I and Mazarin. She even managed to acquire part of the latter's library.[10] Very soon the erudition and aesthetic intuitions of the Queen attracted men of letters such as Isaac Vossius and Claudius Salmasius, from Holland. Even Descartes, in correspondence with the French ambassador at Stockholm, Chanut, came to participate in Queen Christine's newly created Academy.[11] She had realized her intuitions: "Le métier du prince est de faire fleurir les sciences et les arts." (VIII, 39)

b) Her closeness to nature and the fine arts inspired her to engage the most renowned landscape artists in Stockholm: with her, the garden is an allegory of human creativity. The artists, architects, and gardeners she chose were fluent in the Baroque idiom. Nicodemus Tessin and André Mollet designed her French gardens, and the latter dedicated his famous treatise, *Le Jardin de Plaisir*,[12] to Christine's coronation in 1650, to "the model of perfection", in gratitude for her patronage of the expert artists from all Europe whom she had invited to express the magnificence of the Swedish court in their creations. Claude Mollet, the father of André Mollet, is honored in the frontispiece of *Le Jardin de Plaisir*: he had created gardens for Henry the Great, such as St. James Park, and those of Louis XIII – the gardens of Fontainebleau, the Tuileries, St. Germain-en-Laye, etc. The frontispiece[13] of this book is an allegory of achievement: it took several generations to create these magnificent gardens, as buildings, outbuildings, and gardens had to function as a single expression of the magnificence of the master of an estate. According to André Mollet, the four elements (water, air, fire, and earth) also have virtues, it was his deeply held religious belief that they reveal the rational workings of God through nature. While God had distinguished and separated these elements, He nevertheless gave to each an affinity for and participation with the others, seeing to it that they were so disposed that they were able to communicate their virtues together. These gardens, therefore, were intended to symbolize the reconciliation between the elements of fire and water, air and earth, the harmony of man and the cosmos. André Mollet and Christine of Sweden held that the most varied gardens are the most beautiful, as they reflect the diversity and variety of nature. André Mollet proposed to her ceremonial itineraries or promenades in these gardens, to discover the ideal world of harmony, of contemplation, and of the pleasures of the senses. The perfectly orchestrated score of a visitor's walks and stops discloses the garden as a work of art, and an allegory of human creativity in harmony with the cosmos.

While Queen Christine had found both intellectual and aesthetic freedom of expression in Stockholm and Paris, always aspiring to transform the sober characteristics of her surroundings into a world of beauty and harmony, she continued to be inspired by artists whose work is all passion, who by means of their creations had expressed states of the soul as complex and troubled as her own.

IV. PUBLIC OPINION OF QUEEN CHRISTINE

1. *Her Appearance, Her Socially Shocking Observations, and Her Convictions As Found in Her Maximes*

She was drawn to play the part of a man. If the ideal of the *honnête femme* seems to be unattractive to her French contemporaries, it is perhaps because those virtues which are attributed to her would place her in the background of social "divertissements". Mlle. de Montpensier describes Christine in her *Mémoires*[14] of 1656 when she attended a comedy and a ballet with fireworks at the Fontainebleau gardens as having been ill-dressed, ill-behaved, and taking up the attitudes of the two buffoons, Trivelin and Jodelet, one French and the other Italian.

Further, at Compiègne she had asked Madame de Thianges[15] to join her in exile in Rome, as it would be wasted time to live with one's husband, and more, she commanded Mlle. de Montpensier never to get married and to avoid having children.[16] In her maxims she explains the convictions which justify the behavior that provoked the displeasure. Referring to love and marriage, she says:

L'amour & l'hymen sont presque incompatibles (V, 17).
Les hommes s'épousent sans se connoître, & dès qu'ils se connoissent, ils se haïssent (V, 18).
Ce seroit trop heureux, que d'être amoureux et marié (V, 19).

And, on a more positive note:

Il faut admirer le courage de ceux qui se marient (VII, 63), or
Les hommes se marient parce qu'ils ne savent ce qu'ils font, & les filles pour se mettre en liberté à l'abri d'un mari. (VII, 64)

In sum, she realizes the vanity and nothingness of contemporary social conventions and prejudices, which disguise human "nothingness" in relation to a divine Being. "Les hommes rendent rarement justice au mérite et flattent toujours la Fortune" (I, 82). With heroic openness, Christine had rejected the generally accepted social law of marriage and subordination which would have ensured the hereditary succession of the Vasa line. For her, still a Protestant, marriage is not a sacrament and not a religious duty at all. Virtue, according to her, required her to remain immune to the fatal flaw of women who live in subordination and dependence on men. She made most illustrious men enjoy her company in unexpected ways.

2. *Royal and Academic Contacts and Appreciation*

Despite themselves, Louis XIV and the French Academic Conrart were compelled to enjoy her company. Hers was an empire over men's minds. Louis XIV received her at Compiègne, at Chantilly, and in his Louvre apartments: "Le roi, quoique timide en ce temps-là, et nullement savant, s'accommoda si bien de cette princesse hardie, savante et fière, que, dès ce premier instant, ils demeurèrent ensemble avec liberté et agrément de part et d'autre,"[17] reports the Duke of Guise in a letter to Madame de Motteville of 1656. On March 11, 1658 she visited a session of the French Academy, constituted of 40 immortal male members. Conrart reports that she hoped one would forgive her curiosity, obliging Conrart to say that her visit was the greatest honor for the Academy:

> Conrart étoit obligé de dire à sa Majesté que l'Académie françoise n'avoit jamais reçu de plus grand honneur que celui qu'il lui plaisoit de lui faire. A quoi la Reine répondit qu'elle croyoit qu'on pardonneroit à la curiosité d'une fille qui avoit souhaité de se trouver en une compagnie de tant d'honnêtes gens, pour qui elle avoit toujours eu estime est une affection particulière.[18]

V. FROM THE SCIENTIFIC WORLD AND "DIVERTISSEMENTS" TO EXALTED EXISTENCE AND POLYPHONIC SPIRITUAL LIFE: PASCAL, ST. TERESA OF AVILA, QUEEN CHRISTINE

Blaise Pascal, the greatest mathematical genius of the century, sent Christine his calculating machine with a letter filled with wonder and reverence for her absolute sovereignty and sound learning, referring to the accomplishments she had attained in early youth, the excellence of knowledge, and the grace of her womanly nature.[19] Soon afterward, the infinity of the universe was revealed to Pascal by the telescope, and to him, the human soul with its infinite yearnings seemed to echo the heartbeat of the cosmos. Mystical experiences of Pascal and Teresa of Avila were known to Queen Christine. Pascal, on November 11, 1654, in the famous night of fire, had a mystical experience, felt illuminated by grace, and beyond reason, by the intuition of love alone, strove to attain the living and hidden God, the revelation of Whom inundated his heart with joy. Christine did not have either Pascal's night of fire or that exalted experience of Teresa of Avila immortalized by Bernini and Crashaw: St. Teresa's miraculous visitation, the deepest mystical experience, an experience in which she sought to be released from human limitations and be absorbed in the infinite, when Faith in the redemptive power of divine love led to the sudden obliteration of all intervening barriers, traced a path through darkness up to light, leaving all temporal goods, and even reason, far behind.

Queen Christine's quest for divine love is based on intuition. As she seeks the union of the soul with the Divine Spirit through love, a union which can take place in this life and is a prefiguration of the joys of the next, no pain seems to be more intense than that of separation from Divine Love. Christine feels that artists can show her the right path.

In paintings by Rembrandt and Caravaggio, she sees the light of faith about to break through and irradiate the earthbound features of the doubter. When love is beyond the realm of the rational she discovers here a breakthrough of love and the divine spirit. She arrives at the stage Anna-Teresa Tymienecka, in her *Logos and Life* II, *The Three Movements of the Soul*, considers spiritual interiority: "The spiritual phenomenon establishes itself polyphonically: it uses all the fundamental registers that are the powers of the soul: passions, emotions, sensibility, intellectuality, volition, as well as those complex harmonizations which include moral life and creation."[20]

VI. A MAJOR RESOLUTION: A NEW BEGINNING

While both the Stoic philosophers and Descartes had taught Christine self-control and knowledge of herself, she realized that there was a lack of something, and "le fameux connois connois-toi toi-même, dont on a voulu faire la source de la sagesse humaine, n'est que celle de sa misère" (XII, 41). She decided to go beyond the recommendation Descartes gave in the *Traité des passions de l'âme* (1649): "The principal use of prudence or self-control is that it teaches us to be masters of our passions, and to so control and guide them that the evils which they cause are quite bearable, and that we even derive Joy from them all."[21] In her radical self-examination, Christine realizes that there is little liberty in ruling over one of the Great Powers in Europe, and that it is more important to first rule over herself and then to be in contact with all the spiritual forces whose creativity had become the deepest expression of European and international Baroque culture. She is convinced that women should not be rulers: "La loi salique qui exclut les femmes du Trône est juste" (VII, 25).

a) *Her Abdication and Conversion to the Catholic Faith*

Having decided to take residence in Rome and to join the Catholic Church of Rome, Christine abdicated on June 6, 1654, and her regalia and realm were transferred to her cousin, Charles X Augustus. We witness the strength of her conviction in her quest for personal freedom in the following maxim: "Il y a peu de prisonniers plus étroitement gardés que les princes" (I, 23). On Christmas Eve, 1654, she abjured Lutheranism in Brussels. She entered the Catholic Church on November 3, 1655 in Innsbruck, rode into Rome officially on December 23, 1655, and lived there until her death in 1689.

She had left only German and Flemish paintings in Stockholm and had shipped to Rome the Titians, Veroneses, Correggios, and Raphaels from the Rudolph II collection along with an immense library and the most beautiful tapestries.[22] In the Roman *Orbis terrarum*, her life finds a new beginning, a second birth to spiritual values, and Rome too, thanks to her, will enjoy revitalization also, as we shall see. Here she lives according to her conviction: "On doit compter tout le genre humain pour ses parents, & tout l'univers pour sa patrie" (I, 32).

b) *Rome: Christine's Spiritual Rebirth*

Her desire to live in the sacred world of Rome reflected the religious, historical, artistic, aesthetic, spiritual importance of Rome in an age in which both universal peace and Christianity begin to unify paradoxes and disorder into transcendent world harmony. With the defeat of the Holy Roman Empire of the Austrian Hapsburgs, there was now the beginning of the secularized state with religious toleration. Rome had an ally in the Holy Roman Empire, a pan-European, pan-Christian institution based on the *Pax Romana* and the *Lex Romana* of the Roman Empire. To the Romans, God had assigned Rome the task of conquering the world for the spread of universal religion. The idea of a divinely ordained and eternal city of Rome is already pronounced by Jupiter in Virgil's *Aeneid*. Virgil speaks of the Divine Will which leads Aeneas through danger and temptation toward the founding of Rome. Rome also emerges as the ultimate destination of the soul through the *vitae contemplative* in Cristoforo Landino's allegorical interpretation of Rome of 1480.

The arrival of Christine in Rome is a triumph for the pope and was marked by a series of celebrations more splendid than any the Eternal

City had ever known. There she was to live in the Golden Age of theater, ballet, opera, and artistic creation until Pope Innocent XI curbed many cultural activities with spiritual austerity.

Christine expressed herself in a transcendental, ecstatic mood in the following maxims that she feels reflect her experience of the divine illumination which dissolves darkness:

C'est une grande bassesse que de désirer quelque chose de moins que Dieu (XI, 39).
L'âme s'élève en se mettant au-dessus du temps, elle entre dès cette vie dans l'éternité, où tout ce qui n'est pas Dieu disparoit (XI, 52).
Pour écouter Dieu, il faut imposer silence aux créatures (XI, 62).
Dieu parle aux hommes par la nature, par la raison, & par le vérité (XI, 73).
Il faut avoir une grande attention à Dieu, l'admirer, l'adorer, & se laisser faire (XI, 84).
Aimer Dieu & le prochain, c'est la véritable dévotion, tout le reste n'est que grimace (XI, 95).

c) *Rome – Center of Artistic and Spiritual Life: the Quest for the Ultimate Destiny and Significance of Life*

A tenet of Catholic Rome is the spiritual unity of the diversity of experience. First and foremost, Roman gardens, buildings, theaters were created to be enjoyed. From this enjoyment may spring appreciation, and then discernment. Christine was prepared not only to use her intelligence, but also her eyes, and so beheld these artistic creations. Baroque creations are among the most emotional and captivating that art has ever produced. Before these works, she was caught and troubled by their depth, their mobility, their vivacity of expression, and their grandeur. Baroque artists, the most devout since the end of the Middle Ages, had deep personal convictions, and their spiritual life is reflected in their artistic creativity. Rubens, Vouet, Velasquez, Claude Lorrain, Le Brun, Poussin, Zuburán came to Rome. Their art acquired transcendent importance when rival branches of Christendom had to deal not only with new scientific discoveries but with each other as well (the struggle of the Jesuits vs. the Jansenists was especially to the fore in 1656). In Rome, there was an extraordinary rehabilitation of the human will and the rebirth of faith and confidence in the individual, and in the foundation of a genuine world community not in terms of nations or race or culture, but in terms of man in relation to God, to himself and his fellow men and the universe, all of which confirmed the major tenets of Christine's own convictions as a citizen of the world. Queen Christine found in Rome a universe that came to birth from its center, unfolding and stretching out toward the four cardinal points, the four horizons. In the festivities based on ancient rituals, her life found a new beginning.

Other women in the early seventeenth century had travelled as much as Queen Christine. Recall Marie de Medici's journey to Lyons and Paris to marry King Henry IV, that of Anne of Austria to marry Louis XIII and of Margarita who came from Madrid to marry Emperor Leopold I in Vienna. Everywhere they were received with a pomp and ecstasy which overshadowed in festivity all previously given receptions. In each instance the participation of the whole universe was sought, was allegorically represented, in the apotheosis and glorification of the sovereign whose throne was placed at the center of a cosmological totality. In tune with both the cosmic rhythm of renewal and the drama of the Christian cycle, during the darkest days of the year, while the forces of nature slumbered – on December 23, 1655 – Christine found herself at the threshold and boundary between the profane and the sacred of Rome.

We have several engravings that allegorized her arrival in Rome. First, we see her waiting outside the city until the College of Cardinals comes forward to escort her in on December 20, 1655.[23] The engraving is decorated with a garland of sheaves of wheat heavy with the weight of the grain and interlaced with her crown to symbolize both Ceres (Prosperity, Abundance) and Her Majesty.

The next allegorical engraving[24] shows Christina Alexandra (her new name) on horseback with St. Peter's, The Castel Sant' Angelo, and the colonnade in the lower regions of the background. She is very self-confident in realizing a total unity of self, in mind and emotion. Her spiritual virtues guide her to become the foremost patron of art, music, philosophy, and literature. This allegorical portrayal of Christine constructs higher meaning through a metaphoric network of hierarchical analogy and linkage from the human to the divine, from her sensuous desires to her spiritual principles and quest to capture the sacred and ineffable meaning of art and religion.

Then, we see her in a procession sitting in a carriage next to Pope Alexander VI,[25] allegorically on her path toward the center of the world. To her, life is meaningless without an opening toward the transcendent which she found in both the artistic and sacred world of Rome.

The carrousel of *Il Carnavale della Regina* (Feb. 1656)[26] is in tune with the festival spirit as explained above. A mock battle between Amazons (in red and gold) and Cavaliers (in blue and silver), with the two competing teams making four major entrances in four different carriages is staged for the enjoyment of Queen Christine, who sat surrounded by Cardinals Retz, Imperiale, Borromeo, and Azzolini in a box with the best view at the triumphal arch with musicians on its roof, set up in her honor. The four carriages are allegories of Love, Disdain, Virtue, and Time. First, three Graces draw a chariot bearing Roma (*un Amor cieco*), who "in a most majestic manner . . . desirous to discover her joy over the arrival of a Princess so great, seemed to look like love, which being the soul of the World, chose her heart for this unchangeable throne, as a city, the head of the World."[27] Then arrived the chariot of the Amazons drawn by the Furies, carrying "Sdegno," Disdain. The entrance of a dragon vomiting flames with Hercules on its back interrupted the battle between the Amazones and Cavaliers, when Hercules offered the protagonists two apples (apples which are symbols of virtues and the rewards of virtue) which he had stolen from the Garden of the Hesperides. Finally, there arrived the chariot of Phoebus Apollo, drawn by four horses, accompanied by the Year and the four Seasons and twenty-four virgins, allegories of the hours of the day.

With Amor, expressing Rome's love for Christine and the general rejoicing over her decision to reside in the Eternal City, Peace and World Harmony are restored. When seventeenth century writers referred to spectacles, they at first had in mind chiefly the ballets and the operas. The ballet was an expression of universal harmony aiming at restoring political, moral, and musical harmony to the world. According to the Jesuit Père Ménestrier:

> The ancients devised three distinctive styles of dancing: a noble and grave style to express glory and heroism; a free style to represent the various customs to men; and a lively style danced by sylvans and satyrs to represent peace and joy. Each reflects the celestial harmony; each expresses and affects human emotions.[28]

The impact of musical rhythm was felt in the motions of the passions. Mersenne's emphasis on rhythm in his *Harmonie Universelle* (1635) is an indication of the growing importance of ballet. Both Mersenne and Ménestrier clung to the fundamental Platonic concepts of universal harmony and the power of music to inculcate morality through its effects on the emotions. The former believed in the union of music, poetry, dance, and philosophy. Harmony is a divine creation. In creating musical harmony, the composer is providing humanity with the principle of life itself. The dancer was supposed to be the emulator of the planets, guided by the celestial harmony of which the universe was the terrestrial echo. Père Mersenne praised the peacemaking role of measured dancing, the alliance of choreographic figures and dancing. To him, the Creator who regulated the dancing of the stars and of the elements is also the greatest master of the ballet. By means of dance he reckons to teach philosophy, myth, ritual, theology, astronomy, geometry, and architecture. The Jesuit Père Ménestrier sees the ballet as closely related to Christian liturgy, glorifying God in song and dance. He considers the movements from right to left as an analogy to the movement of the sun from sunrise to sunset, a movement he calls "strophes", and those from left to right as movements of the planets, "antistrophes", a pattern we also find in Baroque poetry. He calls "epodes" moments without motion, referring to the impression of the immobility of the earth. The evolution of "strophe" and "antistrophe" signals to Père Ménestrier the evolution of the labyrinth, the allegory of passion as we often find it in literature, ballet, and garden designs. He defines the ballets as actions, customs, and passions "que l'on exprime en ces Dances figurées par cadences harmoniques. Les mouvements sont réglés des gestes, des actions, & des figures."[29] To him, these movements express our passions and feelings and the movements of our soul conjointly.

Festivities in Rome,[30] Paris, and Vienna aimed in principle at the restoration of political and musical harmony in the world. Monarchs, be it Leopold I, Ferdinand II, Louis XIII, or Louis XIV, insured the glory and triumph of their nations through the fine arts, celebrating in festivities the illustrious triumphs of their political victories. They politicized the concept of universal harmony by tying opera, fireworks, banquets to the harmony of the state, by equating morality with loyalty to the monarchy or faith in religion.[31]

While Queen Christine had already patronized the arts and sciences in Sweden, in the manner of monarchs of her age, she chose to continue her activities in Rome in a way even more dynamic than that of Pope Alexander VII, Pope Clement IX and Pope Innocent. These patrons of the arts had engaged the greatest artists of the Baroque age to express faith, grandeur, and obedience and prided themselves with artistic, poetic, and musical creations. Cardinal Rospigliosi, later crowned pontiff under the name of Clement IX, came into contact with the "all the world is a stage" devise of Calderon's theater, and in collaboration with the composers Vergilio Mazzocchi and Marco Marazzoli, he created the Italian *opera buffa* with "Chi soffre speri" in 1639, and "Dal male il bene" in 1653 being based on Calderon's religious work or *auto*, "La vida es sueño," productions in which the characters of the secular drama have been turned into abstractions such as Man, Wisdom, Free Will, Grace, Love, etc. Sigismondo's triumph over himself is likened to the redemption of fallen man through the Sacraments of Baptism and Communion.

VII. THE HARMONY OF SELF AND WORLD IN THE SEARCH FOR TRANSCENDENCE

Christine wholeheartedly inspired and shared the artistic and spiritual life of Baroque Rome. Here she could celebrate the victories which the Church had over heresy and paganism or contemplate the glory of paradise in the creations of illusionistic ceiling paintings in churches and palaces. Foremost, the exuberant vitality, the creative vigor and freedom of expression of Bernini captivated Christine. Bernini, then, was the true interpreter of faith, engaged in ever-mounting festivities in her honor. Bernini's surprise effects were present in both his sculptures and in the dramas he wrote.[32] His artistic creations were a hymn to joy, as Rome was preoccupied with the grandiose and triumphant happiness and lightheartedness of artistic creativity, wherein everything seemed to be in motion toward the transcendent. Nothing was left undone that might draw the beholder into the work of art. The artist wished to captivate the spectator or listener through an appeal to the emotions, and by persuasion at any price. The transmission of emotive experience was the main object of Baroque religious imagery. The rhetorical technique assured easy communication. The majestic ideal of the Church's mission was to inspire the faithful with unreserved admiration, and offer complete satisfaction aesthetically, spiritually, and emotionally, with deep religious intensity. All the arts stood in the service of the religious ideal.

With this in mind let us contemplate the allegory of Christine's glorification, realized in 1689 by B. Fariat after A. Scilla.[33] This allegory expounds the lesson of her life. Here, the varied imagery is inspired by the resurgence of Greek and Roman civilization, seen as an inexhaustible source of legend, moral and philosophical references, a standard for heroism, an aesthetic model for all the arts and literature. Here, the sun exerts its dominion over both all nature and over Christine herself, who knew how to overcome forces of darkness and evil. Like the sun, monarch of nature, she knew how to reject political power. With both crown and scepter at her feet, Christine enthroned is surrounded, from left to right, by Architecture, Study of the Antique, Conversion to Catholicism, History, Sculpture (with an angel standing on the globe and offering Christine a laurel crown with the inscription "Victoria Maxima"), Mathematics, Painting, and Poetry.

At the level of the sun, an angel with palm leaves refers to her spiritual achievements, while cherubs, carrying the heavenly vault on their shoulders, seem to spread Christine's spiritual radiance to a world beyond worlds. All figures depicted in this allegory tend to enlarge space and abolish barriers. Both the marvelous and the real, the invisible and the visible belonged to her imaginative life as she had a lively feeling for the vast realm of the invisible and incomprehensible around her. Hers was the exalted existence that, in A-T. Tymieniecka's words, "frees in man the yearning for the Beautiful, the True, the Just, the Infinite, and the Absolute."[34] Life is not possible without an opening toward the transcendent.

It is in music, above all, that she discovers all the emotions of her innermost being, offering an intuition, a glimpse, or immediate awareness of a reality behind the world of appearances. To her thinking, music breaks through the contingent, mundane world of appearances and refers to a reality which lies beyond or behind it, to a transcendental and sacred reality, both invisible and unexplained.

VIII. THE MUSICAL BEAUTY OF THE WORLD: AN OPENING TOWARD THE TRANSCENDENTAL

Cantata, opera, and oratorio celebrate the musical beauty of the world. Beauty is expressed in eruptive power uprooted by passions. Descartes, in *Compendium Musicae* (1618), had realized that syncopated tones had an exciting effect and emotional depth. Music is the language of the soul and expresses experiences which take place in the secluded depths of the soul. Christine listens to the opera with ecstasy, a language of pure sentiment, liberated from the chains of reason, as she is convinced: "il importe plus de jouïr que de connaître" (I, 35). Listening to opera, she abandoned herself to passion without discussion or reasoning and shared in the pleasure of aesthetic experience without trying to control it. Sorrow and passion are the real revelation of man: "L'ambition et l'amour sont les passions du théâtre" (VI, 59), or "Les passions sont le sel de la vie, qui est insipide sans elles" (IV, 40). Only sufferings and indomitable passions make him what he really is.

The oratorio appeals to her imaginative contemplation. Her favorite composer Giovanni Giacomo Carissimi (1605–1674) concentrated his activity on the Latin oratorio in San Apollinaire where he had been a choirmaster for 46 years. His 145 extant cantatas were much applauded for their "arioso" style which expressed anger and disdain. Vocal music was a popular form of entertainment among members of the nobility. The cantata lent itself to the imagination of the Italian public because of its close relationship to opera and its dramatic elements.

The very structure of the world-soul is a musical one. It rouses in Queen Christine the presentiment of something unexplained and inexplicable in the universe. The impetus of her soul is toward infinity, eternity, and toward boundless knowledge and love. The artistic world is to her a copy of the eternal, and the study of celestial motion and its musical underpinning sheds light on the musical character of the world-soul. Both the modern concert and the modern opera are creations of the Baroque age, whose particular cosmic feeling is reflected in these two parallel expressions of world harmony. Music symbolizes the totality of the world. The world is allegorized as a polyphonic orchestra. The instrument of the world is only complete if no string is missing. The tuned instrument is an allegory of man's soul and the world-soul, an expression of world harmony, when all the senses converge into one harmonious feeling as exemplified in Raphael's "Cecilia" – who is enraptured by heavenly music. Schopenhauer would say, in *Die Welt als Wille und Vorstellung* (I, 3–52), that this painting is the allegory of the transition from the musician to the saint, from the playful beholder of beauty to the rapture of world harmony, a reflection of order, harmony, and peace.

In Cavalli's *La Calisto*, the theme of world harmony is portrayed in dance, drama, and music; the senses range over the entire universe, heaven, earth, the underworld, and even the sea. The action included thunder and lightning, wind, and rain. There were gods in the clouds, monsters in the sea sirens and centaurs on land.

In music, in ceiling paintings, in dance, Christine experiences the Baroque breakthrough of a polyphonic symphony, wherein people surrounded by angels ascend into the air, ceilings dissolve into aerial perspectives, clouds appear through solid vaults, and architecture bursts asunder in fireworks. Music as much as the other arts reveals to Queen Christine a hidden transcendental reality. Music is able to break through the world of appearances, and reveal the inexpressible. Musical revelation is ineffable but knowable. It establishes the foundation of all artistic creations and with them, the miraculous is a natural path of life rooted in the human soul, in the human mind and sensibility.

We have followed Queen Christine's stream of life and may conclude with the words of Tymieniecka: "Life is full: discrete, disharmonious, inconsequential, and nevertheless it is an ever-expanding coherence that surpasses itself at every instance."[35] Illusionism appeals predominantly to the imagination as its magic is in the eye of the beholder, it is poetic illusion in visual form; illusionism is make-believe, very like a theatrical spectacle.

Queen Christine's participation in creativity amounted to her complicity with the artists she commissioned both in Sweden and in Rome to transfigure her life and surroundings as an incentive to her own imagination. The artist's function was to fling open the illusionistic portals into the domain of the imagination, and she entered wholeheartedly. In art, music, and religious experience, she enjoyed a supreme state beyond reason – another world created allegorically in her imagination through the aesthetic perception and embodiment of the sacred. Seeing succeeds where understanding fails!

We hope to have communicated the Queen's confrontation with immense spiritual problems that came as fundamental spiritual values were being steadily undermined by science and philosophy. Aggressive belief in the certitude of reason was badly shaken. Therefore, she turned to the spiritual world of Rome.

IX. ART IS A FUNDAMENTAL WAY OF SEEING: A VISION OF THE WORLD

What is reality for the artist? Artists can deal with visionary reality as something tangible and significant. Merleau-Ponty refers to these special uses of the imagination wherein artists give visible existence to what profane vision believes to be invisible.[36] Profane vision barely reaches the threshold of the sacred, transcendent reality.

Art (ballet, music, sculpture, architecture, etc.) is inseparable from religion and philosophy. The creation of allegories is one of the central ways of ritualizing experience and gaining access to the ineffable, unspeakable, ultimate substance of reality. Implicit in the Baroque vision is allegory, the mystic reality of knowing things by turning into them.

Visions, inspirations, intuitions, the sudden and marvelous sources of the artists' most brilliant creations are life-affirming discoveries in contact with the fundamental life forces. Therefore, the great variety of artistic creations maintain harmony with the four elements which are the basis of everything that exists. Maxims and allegories express a visionary realm, reveal the inexpressible vision experienced with the vividness and power that divine wisdom has deemed necessary to human existence.

With Queen Christine's voyage to Rome, we have experienced the birth of modern Europe, wherein salvation by faith was agreed upon by everyone. In pondering her *Maximes*, we have relived her contemplative life examining allegories we have reactualized both historical and mythical time. There is a transhuman quality to her ways of transcending the human condition. Her existence being open to the world even in her quest for transcendence, there was accomplished in her a veritable ontological mutation.

Michigan State University

NOTES

[1] G. Ph. Harsdörffer, *Lobgesang Dem Hoch-Wohlgeborenem Herrn Carl Gustav Wrangel* and Sigmund von Birken, *Kriegs Ab- und Friedens Einzug* (an allegorical play) quoted in Arthur Lotz, *Das Feuerwerk* (Leipzig: Karl Hiesemann, 1968).

[2] Klaj, "Kastell des Unfriedens", in *Friedensdichtungen und Kleinere poetische Schriften* (Tübingen: Max Niemeyer, 1968), pp. 160–164.

> Der schöne Fried stund schön in solchen Flammen /
> damit nun alle Luft und Freude käm zusammen /
> hat her Fried ein Racquet von seiner Hand gesandt /
> der Zwietrachte Mord Castell zu setzen in dem Brand.

Georg Kohler and Alice Villon-Lechner, *Die Kunst der Verschwendung* (Zurich and Munich: Artemis, 1988); Eberhard Fähler, *Feuerwerke des Barock* (Stuttgart: J. B. Metzlersche Verlagsbuchhandlung, 1974), pp. 16–18.

[3] Henri Chérot, *Etude sur la vie et les oeuvres du père le Moyne* (Paris: 1887), p. 274, note 1.

[4] Engraving. J. Falck Bourdon, "Königliches Schwedisches in Teutschland geführten Krieges" (Stockholm: 1653, Königliche Bibliothek) in "Queen Christine of Sweden, Documents and Studies", *Analecta Reginensia* I, Ed. Magnus von Platen (Stockholm: Kungl. Boktryckeriet P. A. Norstedt & Söner, 1966).

[5] "G. B. Galastruzzi. Final Scene", Il Trionfo della Pieta" in Per Bjurström, *Feast and Theater in Queen Christina's Rome* (Stockholm: Bengtons litografiska A. B., 1966), p. 31 and Engraving. D. Dirickson, 1946, *Analecta Reginensia I.*

[6] "The "Frontispiece" of the Christina Bible", *Analecta Reginensia* I, p. 1.

[7] J. Arckenholtz, *Mémoires pour servir à l'histoire de Christine, reine de Suède*, I, II (Amsterdam, Leipzig, 1751–1760) Vol. II, p. 24. All the quotations in this article are from this edition, all Roman numerals refer to her *Maximes* from this edition.

[8] Sven Stolpe, *Christina of Sweden*, ed. Sir Alec Randall (London: Burns and Oates, 1966), p. 6 and "Minerva", *Analecta Reginensia* I, 1649, p. 337.

[9] Carel van Tug, "The Art Collection of the Teyler Museum". In: *From Michelangelo to Rembrandt: Masterdrawings from the Teyler Museum*, 1909/90, pp. 15–18.

[10] *Ibid.*, and Christine Callmer, "Queen Christina's Library of Printed Books in Rome", *Analecta Reginensia* I.

[11] Arckenholtz, II, *op. cit.*, p. 24.

[12] André Mollet, *Le Jardin de plaisir*. Frontispiece (Stockholm: Henry Kayser, 1651). Jacques Boyceau, *Traité du jardinage selon les raisons de la nature et de l'art* (1638): "God separated opposed elements, but also gave 'à chacun une affinité pour les autres, de telle manière qu'ils puissent communiquer les vertus ensemble.'" Quoted in W. H. Adams. *Les jardins de France, 1500–1800. Le rêve et le pouvoir* (New York: George Braziller, 1980), p. 52.

[13] *Ibid.*

[14] A. Petitot et J. M. Mon-Merqué, *Mlle de Montpensier, Mémoires*, 1656 (Paris: Foucault, 1829).

[15] *Ibid.*

[16] *Ibid.*

[17] *Ibid.* Letter of the Duke of Guise to Madame de Motteville, 1656.

[18] Arckenholtz, I, *op. cit.*, pp. 24–25.

[19] Blaise Pascal, "Lettre à la reine Christine". *Oeuvres Complètes*, ed. J. Chevalier (Paris: Gallimard, 1954): "C'est Votre Majesté, Madame, qui fournit à l'univers cet unique exemple qui lui manquoit; c'est elle en qui la puissance est dispensée par les lumières de la science et la science relevée par l'éclat de l'autorité. C'est cette union si merveilleuse qui fait que comme Votre Majesté ne voit rien qui soit au-dessus de sa puissance, elle ne voit rien aussi qui soit au-dessus de son esprit, et qu'elle sera l'admiration de tous les siècles. Régnez donc, incomparable princesse, d'une manière toute nouvelle; que votre génie vous assujettisse tout ce qui n'est pas soumis à vos armes: régnez donc par le droit de la naissance, par une longue suite d'années, sur tant de triomphantes provinces; mais régnez toujours par la force de votre mérite sur toute l'étendue de la terre."

[20] Anna-Teresa Tymieniecka, *Logos and Life* II (Dordrecht, Boston, London: Kluwer, 1989).

[21] René Descartes, "Traité des Passions". In: *Oeuvres Complètes* (Paris: Gallimard, 1952).

[22] J. H. Elliott, "Queen Christina", *Horizon*, Summer 1967, II, 3, pp. 66–80.

[23] *Ibid.*, p. 73.

[24] *Ibid.*

[25] *Ibid.;* and Naomi Miller, *Heavenly Caves* (New York: George Braziller, 1980).

[26] "Carousel held in Honor of Queen Christina" in the courtyard of Palazzo Barbarini on 20 February, 1656. Painting by F. Gagliardi and F. Lauri in Bjurström, *op. cit.*, p. 62.

[27] *Ibid.*, p. 136.

[28] C. F. Ménestrier, S. J., *Traité des tournois, carrousels, et autres spectacles publics* (Lyon: 1669).

[29] Claude-François Ménestrier, S. J., *Des ballets anciencs et modernes selon les règles du théâtre* (Paris: 1682). Reprint (Geneva: Minkoff, 1972); Marie-Françoise Christout, *Le ballet de cour au XVIIe siècle* (Geneva: Minkoff, 1987).

[30] P. P. Sevin. Watercolor, "Queen Christine at a banquet given by Pope Clement IX on 9 December, 1668", in Bjurström, *op. cit.*, pp. 56–57.

[31] Ernst Léonardy, "Les Fêtes des cours baroques: 'Les Plaisirs de l'Isle Enchantée'" (Versailles: 1664) et "Les festivités à l'occasion du mariage de Léopold Ier" (Vienna, 1666–68), p. 131.

In Vienna there was 'Un tournois de l'air et de l'eau: l'empereur descend des cieux dans le temps des festivités: la dynastie des Habsbourgs a été envoyée sur terre par les dieux (c'est-à-dire la Providence); ce sont également eux qui ont décidé de ce mariage pour apporter aux nations la félicité éternelle.'; p. 145: "L'univers tout entier est intégré dans le fête et participe ainsi à l'hommage; le souverain se voit au centre d'une totalité cosmique. Ainsi dans le premier intermède, les spectateurs virent descendre des cieux successivement de l'Harmonie des sphères (représentation chère aux Florentins), les sirènes, les arques et les Planètes. A la fin du spectacle, Harmonie et Rythme descendent à leur tour sous le règne de Ferdinand."

[32] Gian Lorenzo Bernini, *The Impresario* (Ottawa: Carleton University Center, 1985).

[33] B. Fariat after A. Scilla. Christina sitting on a throne surrounded by, from left to right, Architecture, Study of the Antique, Conversion to Catholicism, History, Sculpture pointing at its work "Victoria Maxima" – the symbol of Christina's abdication –, Mathematics, Painting and Poetry. *Analecta Reginensia* I, p. 4. Compare: "The Glory of San Ignazio – the saint ascending to heaven in a whirlwind" or Pietro da Cortona and Tymieniecka, "The Glory of the Barberini" (Palazzo Barberini, Rome).

[34] *Logos and Life*, II.

[35] Tymieniecka, *op. cit.*, XXVII: "Life is full: discrete, disharmonious, seemingly inconsequential, and, nevertheless, it is an ever-expanding creative coherence that surpasses itself at every instant."

See also: Francesco Cozza: allegory of Music and other ceiling figures – ceiling fresco in the Library of Pamphili, Collegio Innozenziano.

[36] Maurice Merleau-Ponty, *L'Oeil et l'Esprit* (Paris: Gallimard, 1963).

Fig. 1. Engraving. D. Dirickson. Queen Christine.

Fig. 2. Arrival in Rome, unknown artist. J. H. Elliott. "Queen Christina". *Horizon*. Summer 1967, II, 3.

Fig. 3. Painting by Gagliardi and F. Lauri. "Carrousel held in Honor of Queen Christina" in the courtyard of Palazzo Barbarini on 20 February, 1656. In Per Bjurström.

Fig. 4. B. Fariat after A. Scilla. Christina sitting on a throne surrounded by, from left to right, Architecture, Study of the Antique, Conversion to Catholicism, History, Sculpture pointing at its work "Victoria Maxima" – the symbol of Christina's abdication –, Mathematics, Painting and Poetry.

JADWIGA SMITH

ALLEGORY AND THE DRAMATIC FOUNDATIONS OF THE MEDIEVAL THEATER

The investigation of the dramatic principles of mystery or morality plays is not a novel task. Jerome Taylor's "The Dramatic Structure of the Middle European Corpus Christi, or Cycle, Plays," Richard J. Collier's *Poetry and Drama in the York Corpus Christi Play*, and Robert Potter's *The English Morality Play: Origins, History, and Influence of a Dramatic Tradition* are just a few examples of the critical interest in the dramatic aspects of the medieval drama, whether associated with staging, the function of poetry, plot structure, or ideological content. My investigation of the dramatic will focus on the role of allegory in creating both the artistic unity and dramatic tension necessary to sustain the attention of the audience. I do not assume that allegory is the sole foundation of the medieval drama, nor do I try to ignore the fact that allegory has quite obviously been studied in the case of morality plays. Instead, I propose to focus on the role of allegory in the creation of a certain dramatic tension by means of the *via negativa*. In other words, establishing the reality of a given play and its spiritual impact by means of a skillfully applied interplay between allegorical representation of the obviously true and the allegorical representation of the obviously false adds dramatic quality to the medieval plays.

The concept of negation is, of course, dangerously associated with that of falsehood. But they are not the same even though it may happen that the negation is true and the corresponding affirmation is false. But in denying a statement, we implicitly or explicitly assert that the statement in question is false, although, of course, the assertion that something is false may itself be true. Also, when a proposition is complex, it is often important to distinguish the negation of the proposition as a whole from propositions resulting from the negation of some component or components of it. Thus, in short, by the careful application of negation, we can achieve confirmation and, also, by asserting the truth of a large whole, its components will be more striking as true when put together with false components – those only pretending to represent the whole.

The whole world in the case of the medieval drama is the entire world, the cosmos with God in charge. This "true" world, founded on God's

M. Kronegger and A-T. Tymieniecka (eds.), Analecta Husserliana XLII, pp. 29–35.

principles, is beyond doubt and question. Still, its "truth" has to be communicated and taught to the believers in a perpetual cycle of teachers and pupils. At the same time, a mere recital of the articles of Christian faith would not sustain the interest of pupils, or, in this particular case, the audience. In other words, in addition to the practical, that is the pedagogical aspect of theatrical performance, there is also a need for a dramatically rendered feeling of awe, a need to express a grand, almost everlasting sense of magnitude of what is to be revealed through the medium of the stage play. Thus, the battle emerges between the pedestrian and the sublime, between the lowly and the holy.

These opposites, however, are not involved in tight plot structures according to Aristotelian principles that build up conflicts and strifes around clearly delineated protagonists and antagonists. This is to say, the medieval stage characters do not participate in a cause-and-effect chain of events leading to a carefully prepared and orchestrated cathartic culmination. Granted, the non-Aristotelian foundations of the medieval theatre still, obviously, include various aspects of plot development and stage organization, but the principles of contrast and inversion seem to dictate the most basic dramatic frame for the medieval plays. Thus, the cause-and-effect in a mystery play will be outside of it because a particular play is only one link in a chain of events involving multiple plays in order to represent the history of the world, starting with its creation.

Thus, from their connections in the mystery plays to the psychomachian struggles in morality plays, it is through inversion that the truth of faith is to be revealed without, on the one hand, boredom and the danger of the blasphemy of false assuredness. On the other hand, the *via negativa* is inherently dramatic because by avoiding a static, passive acceptance of the components of faith, it challenges the audience to participate actively in unravelling the very fundamentals of their spirituality.

The role of allegory, then, in the medieval drama is intended both to help with the instruction in fundamentals of faith and to coax into a spiritual probe. The results are obviously to be those of a successful lesson as well as a test. That allegory in the medieval drama would assume such a role was a result, on the one hand, of the tradition following Prudentius' *Psychomachia*, in which highly polarized moral allegories are eventually involved in a violent physical battle. On the other hand, another tradition, resulting from the affinity of the medieval public for debates, trials, and disputes called, quite naturally, for a clear representation of opposing views and concepts by means of allegory. There are numerous examples of these debates and trials in the medieval drama. The final trial scene, for instance, in *The Castle of Perseverance* with a long debate among the four virtues is just one of these examples. The addition of the atmosphere of testing and probing comes not only from the tradition of debates in which the audience would take sides but also from the fact that allegory would challenge the audience's knowledge of the theological and spiritual cause. Such testing would result from the requirement, on the part of the audience, of having a prior knowledge of the meaning of a given allegory. The character of World, for example, in *Mary Magdalen* from the Digby Manuscript seems to be charming and impressive, but World's true hideous identity of a tempter is known all along. In the same manner, the Vices in the play advertise their distinctive attributes without any shame, but the audience knows better and almost automatically reverses those qualities into their opposites. By the same token, the audience is aware of the constant threat to the Virtues, even at the time of their apparent triumphs. The attractive sins eventually show their true colors, but by then the audience manages both to enjoy and disapprove of the situation of temptation. As a result, allegorical representations on stage call for simultaneous negations. For example, Human Genus in *The Castle of Perseverence* refuses the sensible service of Bonus Angelicus only to follow Malus Angelicus into a life of sin. One of the sins, Flesh (Caro), talks sweetly:

> I am Mankindys faire Flesh, florchyd in flowrys;
> My life is with lustys and likinge i-lent.
> With tapitys of tafata I timbyr my towrys;
> In mirthe and in melodye my mende is i-ment.
>
> (Bevington 807, lines 237–40)

Thus, to respond properly to allegorical representations by means of negation is part of a dramatic tension resulting from testing the ability of the audience to recognize the inherent unity of opposites.

This *via negativa* method as a means of coping with the challenge of expressing highly charged spiritual and emotional issues from a long tradition of exploring the possibilities of paradox in the area of theology, whether to confirm or challenge the orthodoxy. To grasp the essence of the transcendent deity, theologians and philosophers often have used paradox, especially through the application of the logical and rhetorical, as well as the epistemological negation of various relegious questions. For example, the *via negativa* method was considered by Dionysius to be:

more appropriate to the divine essence, since, as the secret and sacerdotal tradition thought, we rightly describe its non-relationship to things created, but we do not know its superessential, and inconceivable, and inutterable indefinability. If, then, the negations respecting things Divine and true, but the affirmations are inharmonious, the revelation as regards things invisible, though dissimilar representations, is more appropriate to the hiddenness of things unutterable. (Part 2, 7–8)

Also, negation helps to avoid lying about God since its does not provide the statement of what God is, but of what He is not. As a result:

Divine things should be honored by the true negations, and by comparisons with the lowest things, which are diverse from their proper resemblance. There is then nothing absurd if they depict even the Heavenly Beings under incongruous dissimilar similitudes, for causes aforesaid. (Part 2, 7–8)

The growing acceptance of scholasticism also brought a renewed interest in Aristotle and his logic. Thus his treatise *On Interpretation* provided an insight into the application of negation:

Since it is possible to enunciate that what belongs to a subject does not belong to it and what does not, does, and what does not, does not, and to enunciate these in regard to those times outside of the present as well as of the present, it could be possible to deny whatever someone affirms and to affirm what he denies. It is evident, therefore, that there is a negation opposed to every affirmation opposed to every negation. (73)

In this period of mounting religious disputes, Christian belief found itself to be challenged, especially intellectually, by a more and more acknowledged awareness of the paradoxical nature of a number of essential articles of faith. Medieval allegory thrived on such a challenge, however. By trying to curb the abstract, allegory had applied a direct or even head-on approach to theological and spiritual concerns of the Christian faith, from the concept of free will to the sacraments.

Of course, the *via negativa* plays a dominant role in medieval drama as an obvious source of dramatic tension in the mystery plays. By means of multi-layered typological interpretation, the mystery plays gain both in the dramatic unity of a whole cycle as well as in the dramatic tension of a particular play. For example, Noah's wife from the Wakefield Corpus Christi Cycle is to be seen in a typological relationship with both Eve and the Virgin Mary. Noah's wife's lack of humility and her refusal to understand her husband's mission are in contrast to Mary's virtues. In other words, we are witnessing a case of the *via negativa* in order to express Mary's spiritual morals to a multiplied degree. At the same time, the audience, forced to pass such a judgment, has to make all the proper connections using various plays within the cycle as well as in the outside context of their religious knowledge and spirituality. The resulting tension is clearly dramatic not only because of its emotional impact but also because of the connections made by the audience, allowing the audience to justify the place of a particular episode within a larger scheme of events. Thus, any dramatic "loose ends" are avoided.

In the case of the passion plays the *via negativa* emphasizes Christ's humanity by the lack of humility on the part of his tormentors. In the Wakefield Scourging, Christ remains silent throughout his suffering and thus acquires the status of the allegorical representation of humility. In other words, the meaning of the silent figure on stage emerges from the doctrinal knowledge on the part of the audience and not from the direct meaning of the blasphemous words spoken by the tormentors. Rather, the impact of their speeches and the excess of their cruelty require a similar reaction on the part of the audience to that evoked by Vices. Their speeches and behavior, of course, are often contrasted with subdued and meek Virtues. The dramatic effect, then, in passion plays comes from such a contrast.

Actually, the whole medieval drama exhibits intense polarization, the fight between good and evil, and knowledge and the lack of knowledge. The mystery cycles provide the vision of history of the world as constant battles between the opposites, oftentimes reinforcing the results of tense conflicts by the use of negation.

The morality plays perfected the use of the *via negativa*. As Robert Potter observes on the sinful acts of man in the morality plays:

> The irony of this course of action (and no doubt the source of much dramatic appeal for a medieval audience) lies in the mockery of established values which the logic of the drama presents. If all men are sinners, then man's rejection of sobriety and good order must be presented as sensible and at least, temporarily inevitable. Virtue, for the moment, must appear barren and static; freedom must lead to sin, in the sense of pleasure and with the reinforcing implications of fertility. The protestations of reason (in the figure of Mary, for example, in *Mankind*) must for the obligatory moment, be overruled by misrule (as personified in the figures Mischief and New Guise in the same play). (34)

As mentioned earlier, the audience can recognize its own vices: The audience is able to perceive, through reason and moral conscience, the inherent paradox that the obligatory happy end of a morality play is only the result of the threat to the world's *status quo*.

Then, the audience gets the truth, which comes with the victory of the virtues on the stage. At the same time:

> The audience drawn together on a religious and communal occasion to be entertained and frightened by the caricature of its own behavior, is nevertheless verifying fundamental Christian concepts; each of the plays is an act of faith in life, considered as a process. The human predicament is imagined not a static circumstance, but a linked sequence of events – not as a portrait, but as a drama. (Potter 36)

We can say, then, that at one point virtue has to appear weak and vice enchanting. But since the outside source of reference prevents confusion, the awareness of the underlying principle of negation is a source of the dramatic tension.

Thus, in conclusion, the medieval allegory thrives as a dramatic medium of conveying the otherwise boring pedagological exegesis of the doctrinal, spiritual, and theological foundations of faith. The dramatic principles of negation, contrast, and properly discussed distortion are responsible for making allegory one of the essentially inherent dramatic devices. The essentially inherent is to be understood as not belonging to theatrical embellishment such as scenery, music, acting, that is of supplementary value. In the absence of a tight plot or well-developed characters, as well as lengthy speeches with little consideration of stage time, allegory helps sustain the audience's attention with a constant appeal to participate in an almost game-like activity of unravelling the dramatic fusion of opposites essential to allegorical representation.

The simultaneous awareness, however, that an outside order firmly exists, providing points of reference for all the distortions of the real world on stage, is an essential factor allowing the audience to process the meaning of the world represented on stage. But, with the passage into the Reformation, many of these points of reference will be lost, and negation becomes a vehicle for expressing growing doubt in order – whether social or religious. From that time on, the allegorical pairing of negations, direct or implied, is no longer important in constituting the dramatic foundation of the post-medieval theatre.

Allegory gives way to metaphor. The symbolic and ironic representation of Renaissance drama focuses on multiple connotations. At the same time, the dramatic foundations of the new drama fall more into the Aristotelian mould – with clearly defined characters within one and the same play, well-developed plot lines, and relatively self-contained represented reality; that is, the cyclical drama disappears. The outside context, in the absence of the belief in an external order, becomes a source of multiple allusions rather than a supply of certainty. The principle of negation with its polarization of opposites is now too simple for the purpose of the Renaissance drama, which, instead, blends and blurs the opposites. Also, irony is no longer based on the principle of a simple inversion; rather, it employs multiple and often opaque associations.

Contrasts and opposites become, as a result, contradictions leading more toward relativism than to certitudes. By the time Jacobean drama emerges in the midst of religious, political, and social discord, the drive towards metaphor, symbol, and irony seems to be accomplished.

Bridgewater State College

REFERENCES

Aristotle, *On Interpretations*, trans. Jean T. Oesterle, Milwaukee: Marquette UP, 1962.

Bevington, David, *Medieval Drama,* Boston: Houghton Miflin, 1975.

Dionysius the Areopagite, *On the Heavenly Hierarchy: Works,* trans. John Parker, London, 1982.

Potter, Robert, *The English Morality Play: Origins, History, and Influence of a Dramatic Tradition,* London: Routledge and Kegan, 1975.

CHRISTINE RAFFINI

BALZAC'S ALLEGORIES OF ENERGY IN *LA COMÉDIE HUMAINE*

Le vieillard se tenait debout, immobile, inébranlable comme une étoile au milieu d'un nuage de lumière. Ses yeux verts, pleins de je ne sais quelle malice calme, semblaient éclairer le monde moral comme sa lampe illuminait ce cabinet mystérieux.

Balzac, *La Peau de chagrin*

"We cannot speak," C.S. Lewis observes, "perhaps we can hardly think, of an 'inner conflict' without a metaphor; and every metaphor is an allegory in little. And as the conflict becomes more and more important, it is inevitable that these metaphors should expand and coalesce, and finally turn into the fully-fledged allegorical poem."[1] Indeed, allegory may seem as universal as conflict itself, and allegorical poems flourish well beyond the Middle Ages, although often in surprisingly altered forms. If allegory can be defined in a general way as the personification of abstract concepts, we might add that it does not disappear; even as late as the nineteenth century it can be seen to find a new and vital expression in the vast poem, which Honorè de Balzac – with Dante obviously in mind – named *La Comédie humaine*. Like Dante, many medieval poets are predisposed toward experience " . . . of personified beings of a supersensual nature . . . "[2] and Balzac as well, through his interest in esoteric lore, often inclines naturally toward allegorical modes of expression. Moreover, in late antiquity and in the Middle Ages, allegory served to bring " . . . poetry close to philosophy . . . " – a rapprochement which Balzac also achieves in his philosophical novels.[3]

In medieval allegory, Love, Wrath, Clemency – indeed, almost any abstraction might be personified; in Balzac's world instead we discover, as I now hope to show, only two opposing forces which combine and recombine in a surprising number of variations. These expressions can be good or evil, positive or negative according to the context in which they are found, and thus they are in many respects reminiscent of medieval psychomachia. However, unlike medieval allegories which tend to consist of one-dimensional and easily identifiable embodiments of

M. Kronegger and A-T. Tymieniecka (eds.), Analecta Husserliana XLII, pp. 37–47.

virtues and vices, Balzac's personifications are first of all complex characters, replete with habits, desires, aversions and ambitions. In writing off these characters as inflexibly and rigidly constructed on a particular central passion, many critics have failed to take this complexity into account. In the following pages, I will attempt to show that many of the protagonists of *La Comédie humaine*, despite their great diversity and complexity, consistently personify on the one hand forces of conservation, preservation, retention and expansion and, on the other, opposing forces of release, self-sacrifice, prodigality and dispersion.

In contradistinction to traditional allegory, these forces are personified to the extent that they essentially determine, motivate and galvanize the characters in question. For example, greed, a negative aspect of the force of conservation, governs and determines every aspect of the thinking and behavior of a secondary character such as Madame Vauquer. Clearly, this abstraction is the overriding force in her 'existence'. In that she is inconceivable without it, does not in fact think a thought or utter a word that is not motivated by it, she can be said to personify it. In a somewhat more complex fashion Balzac's major characters also develop, evolving and fulfilling their destinies as virtual personifications of either forces of conservation or of the opposing forces of dispersion.

In certain instances, these antithetic forces combine harmoniously in one individual. Where, for example, friendship, loyalty and love are involved, characters preserve their existence and define themselves by giving and even sacrificing themselves. Thus Henriette de Mortsauf, the heroine of *Le Lys dans la vallée*, knows that the infinite care and attention she bestows on her two sickly children is tantamount to giving birth to them every day: "I will blow life into them," she promises herself. I will give birth to them every day."[4] Through this perpetually renewed effort she literally keeps them alive, sustaining as well her own strength and voluptuous beauty. When Félix de Vandenesse, however, causes her to fall in love, the symbiotic balance that preserves her and her children is irreparably disturbed. Guilt, frustrated desire and jealousy finally vitiate the life-giving transfer of energy, causing her to fall ill, waste away and die, as her doctor explains, of hunger: "Rich, young, beautiful, and dying emaciated, aged by hunger, for she will die of hunger! For forty days, her stomach, as if it were closed, has rejected all nourishment in whatever form it is presented."[5] Likewise, Henriette, in the name of virtue, has rejected Félix's advances while refusing to relinquish the thought of him. When he arrives at last at her side, he finds her hideously changed, for she has become an outward manifestation of the hunger that comsumes her:

She was no longer my luscious Henriette, nor was she the sublime and saintly Madame de Mortsauf, but instead Bossuet's nameless thing which struggled against the void, and which hunger and thwarted desire had pushed into the egotistical combat of life against death. . . . Her colorless lips spread across her hungry teeth, attempting one of those forced smiles in which we also hide the irony of vengeance, the expectation of pleasure, the drunkenness of the soul and the fury of disappointment.[6]

Madame de Mortsauf, who lives and dies for virtue as well as for love, indeed exemplifies self-sacrifice. Old Goriot, however, who, to the very end, draws his vitality and the whole meaning of his life from dwelling on the thought of his absent daughters, embodies perhaps more completely than she the force of self-sacrifice. His parental love amounts to veneration:

So then! When I became a father, I understood God. He is everywhere, completely so, since creation came out of him. Sir, I'm like that with my daughters. Only I love my daughters more than God loves the world, because the world is not as beautiful as God and because my daughters are more beautiful than I.[7]

Unlike Henriette de Mortsauf, who was able to give her children love and attention, and ultimately to justify herself to Félix, old Goriot, because of his daughters' selfish disdain, can only worship them from afar as he doles out his worldly goods to them and their husbands. When the supply is at last exhausted, he also is consumed and dies.

Parental love is a recurrent theme in this novel: the same life-giving tendency also motivates the noble and refined Vicomtesse de Beauséant as well as the criminal, Vautrin, both of whom act as mentors and teachers to the young Eugène de Rastignac. When she explains the sordid nature of Parisian society to the young man, her disinterested advice is as exact and compelling as Vautrin's: if one is not the executioner, she warns, one will become the victim. Both, the Vicomtesse and the ex-convict, despite their differing motives, speak to Rastignac with the same probity and passionate intensity. Although Vautrin's intentions are to exploit and seduce Rastignac, there is no duplicity in his apparent means. In connection with Lucien de Rubempré he will, instead, resort to duplicity, but in the end, a forthright expression of paternal devotion characterizes his relationship with the young man. Upon learning of Lucien's suicide, he is dumbfounded:

This blow for me is much worse than death, but you can't know what I'm saying. . . . You're a father, if you are one, only in one sense; . . . I am a mother. too! . . . I . . . I am crazy . . . I feel it.[8]

Lucien is indeed Vautrin's creation. Like all the sincere pedagogues, devoted friends and faithful servants in *La Comédie humaine*, Vautrin, the outlaw, the corrupter and tempter of youth, also embodies the energy of self-sacrifice as well as that of self-renewal and preservation.

Lisbeth Fischer, the hard and cunning protagonist of *La Cousine Bette*, instead only simulates self-sacrifice. While methodically eroding the foundation of her relatives' lives, she, in their eyes, remains a charitable figure – part of the family. Driven by her thirst for revenge, she orchestrates their ruin while cleverly preserving her own reputation. She works patiently yet swiftly in order to demolish:

. . . this family which every day grew odious to her, for one hates more and more, just as one loves more every day, when one loves. Love and hatred are feelings which feed themselves; but, of the two, hatred lasts the longest. Love is confined by limited forces, receiving its powers from life and prodigality; hatred resembles death, greed, and is in a sense an active abstraction, above beings and things.[9]

Bette is indeed the embodiment of hatred, and although in a sense blinded by it, she nonetheless succeeds in skillfully directing it to her own ends. Her machinations are despicable, but highly plausible. A critic notes that "A full decade before Freud was even born, Balzac has endowed this poor relative from the provinces with a psychological depth that anticipates all analytical studies that were yet to come. He speculates with her character on the lasting effects of early trauma and the crippling force of unresolved emotions."[10] Balzac specifically attributes her resourcefulness to her primitive nature, alluding as well as to her virginity. "When chaste people need their bodies or their souls, having recourse either to action or to thought, they then find steel in their muscles or science infused in their intelligence, a diabolical force or the black magic of willpower."[11] As Bette preserves her virginity, she preserves to her death the secret of her jealous loathing.

Like Bette, like Vautrin, all those who cultivate patience and who renounce immediate gratification are those who attain their sought-after goals. Grandet, for example, possesses extraordinary patience. This rapacious and vigilant miser, from whom nothing escapes, who lives in order to weigh, brood, and retain not only gold, but gestures, words, even syllables, always measures at every step the breadth of his grasp and power, so that each investment, no matter how seemingly inconsequential, might bear fruit. Even in death he is not destroyed: his obdurate discipline, his powers of concentration and his fortune are perpetuated in his daughter, Eugénie, who, although her father's nemesis, is endowed with his stubbornness and his will. Unlike her father, however, Eugénie is by nature a loving, giving being: "faite pour aimer." Ultimately forsaken, unable to fulfill her natural destiny, she accepts the monotonous patterns of existence established by her father, without, however, being untrue to herself or uncharitable toward others. At forty, she is exceedingly rich, but, still obedient to her father's wishes, lights a fire in her room only on those days when he would have allowed it.

> The house at Saumur, the house without sunlight, without heat, incessantly in shadows, melancholy, is a mirror image of her life. She carefully accumulates her revenue, and perhaps would seem parsimonious if she didn't belie gossip through the noble use of her fortune. . . .The hand of this woman bandages the secret wounds of every family. Eugénie walks to heaven accompanied by a cortège of good deeds. The greatness of her soul lessens the narrowness of her upbringing and the customs of her early life.[12]

Thus like her father, Eugénie exemplifies the force of conservation, but tempered by her purity and goodness, it is manifest as a redeeming strength.

In *La Comédie humaine* the force of conservation when carried to extremes can be far more destructive than any impulse toward dissipation. The latter, for example, in the case of Baron Hulot, enslaves the senses and the will, but the former has the power to annihilate even the spiritual dimension of being, and it is precisely these circumstances which form the conflict in *Louis Lambert*. From childhood, Lambert, the misunderstood genius, is gifted with total recall. He understands everything:

> A man of ideas, he had to slake the thirst of his brain which sought to assimilate all ideas. Whence his readings, and, as a result of his readings, his reflections which gave him the power to reduce things to their simplest expression, to absorb them into himself to study them there in their essence. . . . In order to exist didn't he have to toss fodder incessantly into the abyss he had opened in himself?[13]

Even the metaphor suggests the vast reserves of energy needed to sustain this character's life. Indeed, Lambert does not study – he devours knowledge and ideas, incorporating them and preserving them in himself, and through him Balzac explores his own notions of the materiality of thought and will, drawing on notions gleaned from reading Lavater, Mesmer and Swedenborg. There are, according to Lambert, three levels or spheres of being: the lowest is assigned to instinct and feeling, the second to reason and abstract thought, and the highest to sight and ethereal vision. Lambert, however, despite his genius, creates nothing, produces nothing save a single book, "Le Traité sur la volonté," which is lost before anyone can read it. In the beginning he attempts to explain it, but, finally, having understood and felt everything – having attained omniscience – he has nothing left to say, and, incapable of communicating what he knows, he is cut off from the world. He is not, however, invulnerable to love: "As soon as he noticed Mademoiselle de Villenoix, he sensed the angel within her form. The rich faculties of his soul, his inclination toward ecstasy, everything in him was then resolved by a limitless love. . . . "[14] But the power of love, rather than reconnecting him to life, threatens to annihilate the vast stronghold of his inner self. To avoid physical involvement, he escapes at the last moment into insanity, becoming the most unsettling and perplexing example of the force of conservation in the work of Balzac. Seeing him after the crisis, the anonymous narrator describes him as " . . . a debris snatched from the grave, a kind of conquest made by life over death."[15] Mademoiselle de Villenoix instead affirms that he is in heaven, gifted with ethereal vision, yet still conscious of the world, although perceiving it in a form that is incomprehensible to mortals.

Raphaël de Valentin's fate in *La Peau de chagrin* both complements and opposes that of Louis Lambert. Both protagonists are the victims of occult forces, and, significantly, both compose treatises on willpower. Both are in love with women named Pauline, but there the resemblances seem to end. Whereas Lambert's struggle illustrates the ravages of conservation, Raphaël de Valentin's emphasizes those of waste. From the outset, Raphaël is presented as a fallen creature, his name further heightening both the allegorical and ironic overtones of the story. The life of this lost "angel without rays" unfolds under the aegis of dissipation: having gambled away everything in a single night of debauchery, he resolves to commit suicide. While waiting for nightfall, he wander into an antique shop, and there amid the accumulated debris of "twenty worlds" he finally notices a piece of shagreen, no larger than a fox skin, but unusual in that it casts an inexplicable glow in the darkness of the shop.

As the new owner of this magic talisman, Raphaël soon discovers that under its spell, each uttered wish saps his energy and shortens his life, degree by measurable degree. Ironically, he who once lavishly squandered himself in the pursuit of pleasure, and then in despair sought only to put an end to his life, is finally forced to conserve at all costs a colorless and desireless existence. His servant Jonathas has been instructed to anticipate Raphaël's every need:

I must give him his dressing gown, always made up in the same fashion and of the same material. I am obliged to replace it when it wears out, just to spare him the trouble of asking for a new one. . . .The menu for the whole year is made up in advance, day by day. Monsieur le marquis has nothing to wish for. He has strawberries when there are strawberries, and the first mackerel to arrive in Paris, he eats it.[16]

This painfully contrived mode of existence contrasts with the positive self-preservation which first characterizes both the cold and haughty Foedora and the mysterious keeper of the antique shop. The latter attributes his remarkable longevity directly to the powers of seeing and knowing. Likewise, Vautrin and all those in *La Comédie humaine* who succeed as a result of a single-minded drive toward conservation, value seeing, watching and knowing. As Vautrin explains: "I have obtained everything because I knew how to disdain everything. My only ambition was to see."[17] Raphaël, on the other hand, as the embodiment of desire, is destined to wear himself out. The old shopkeeper warned him of the danger:

The soul is composed of terrible poisons as a result of the rapid concentration of its pleasures, through the strength of its ideas. . . . Wanting burns us up and being able destroy us: but knowing leaves our weak organization in a perpetual state of calm.[18]

As he consumes himself, Raphaël becomes a virtual allegory of the power of these two verbs, *vouloir* and *pouvior*. In this respect, he becomes a paradigm of the young Balzacian protagonists who, spurred by ambition and goaded by desire, risk everything. If, however, they, like Rastignac or especially like Vandenesse, learn to think coolly and to observe and listen patiently, their success is assured.

Lucien Chardon de Rubempré despite his aspirations, stands apart from the other protagonists. Like Rastignac and Vandenesse, he is young, ambitious, good-looking and of provincial origin, yet he lacks their perseverance. He tends toward profligacy, yet without being driven by the will to pleasure that characterizes Raphaël. Although he clearly belongs in the rank of the squanderers, his case is more complex. He evinces the simplicity and the undisguised selfishness of a child. His bitterest disappointments can be assuaged by the merest distraction: a meal in a restaurant, for example, has the power to soothe his wounded vanity. Among the handsome, young Balzacian protagonists, Lucien alone proclaims himself to be a poet, yet, entirely lacking the tenacity and concentration to express his artistic vision, he instead flutters from attraction to attraction, continually shifting his alliances and scattering his energy. Thus divided, leading several lives at once, he seems both less and more than a single entity. In prison, he must at last confront the reality of his shameful failure, compounded, it seems, by his multiple selves:

. . . suffering from all his overturned hopes, hurt in all his social pretentions, crushed in his annihilated pride, in all the selves presented by the ambitious one, the lover, the fortunate one, the dandy, the Parisian, the poet, the voluptuous one and the privileged one. Everything in him had broken in this Icarian fall.[19]

Balzac often chooses to stop short of omniscience in rendering this character, and thus the reader sees primarily an exterior, a physical presence of extraordinary beauty given to involuntary smiles, clever quips and abundant tears. In prison, we are privy to Vautrin's monologue, but to the end, Lucien remains an impenetrable mask. When at last he prepares to kill himself, his divided nature is once again accentuated:

Lucien saw the palace in its original beauty. . . .He accepted this sublime view as a poetic adieu from civilized creation. While making calculations for his death, he wondered how this marvel could exist unknown in Paris. There were two Luciens, Lucien, the poet, taking a walk in the Middle Ages, under the arcades and towers of Saint Louis, and a Lucien preparing to commit suicide.[20]

Because of the young man's nature, Vautrin is able to live through him; Lucien becomes in fact Trompe-la-Mort's alter ego:

His powerful faculties, absorbed in Lucien, functioned only for Lucien; he took delight in his progress, in his loves, in his ambition. For him Lucien was his visible soul. Trompe-la-Mort would dine at the Grandlieu's, slip into the boudoirs of great ladies, love Esther by proxy. In a word, he saw in Lucien a Jacques Collin who was handsome, young, noble, on his way to becoming an ambassador. Trompe-la-Mort had realized the German superstition of the DOUBLE through the phenomenon of moral paternity.[21]

Indeed, Lucien is everyone's *doppelgänger*: a long sought-after reflection, a mirror, a soulmate, a kind of nexus of powerful transfers of energy.

Careless squanderer, wastrel, ingrate, he exploits others just as easily as he falls prey to each temptation that comes his way, afterwards weeping hot tears of repentance. Curiously enough, deprivation, humiliation, and failure do not lessen him; up to the end, he remains invincible, thus appearing to defy the Balzacian system where endurance is determined by the amount of inner strength and willpower a character possesses. As we have seen, Lucien has neither one. He is, in fact, a kind of mirage, which may explain the tendency of other characters to seek to live through him. Not only Vautrin, but Lucien's friends, lovers and family invest all their hope, strength and even their fortunes in him. When he first leaves home, his family continues to support him: " . . . Eve, Madame Chardon and David had offered to the poet, each of them separately, the purest of their blood."[22] As Daniel D'Arthez, the high-minded friend whom Lucien has betrayed, explains in a letter to Eve:

> Rest assured, Lucien will never go as far as crime – he wouldn't have the strength, but he would accept a crime that had already been committed. He would share the profits without having shared the dangers, which seems horrible to everyone, even to criminals. . . . Lucien is a harp whose cords grow taut or slack in response to atmospheric variations. He might write a beautiful book in a phase of anger or in one of happiness and not be sensitive to its success – after having nonetheless desired it. . . . He is a brilliant assemblage of fine qualities embroidered on a canvas that is too flimsy; age carries off the flowers, and one day only the fabric remains, and if it is of bad quality, one sees only a tattered rag.[23]

Lucien, then, perhaps more than any other character, embodies the force of dissipation in that the vitality and substance he squanders are never his own, but are drawn from those around him, and whereas they, with the exception of Vautrin, are diminished, he continues to thrive, untouched by life. Even in death, he seems to lack substance; in prison, friends are shocked to discover: " . . . Lucien hanging as if his clothes had been placed on a porte-manteau. . . . "[24] An insatiable void during his lifetime, even after his death, Lucien continues to sap the strength and hope of those who knew him. Camisot, the judge, assesses his ruined career, noting that the dead have power over the living: "Lucien carries our hopes into his coffin."[25]

Thus Lucien de Rubempré, in projecting the most negative aspects of the force of dissipation, strikingly illustrates another aspect of Balzac's use of the allegories of energy. He is exceptional in that his energy derives almost exclusively from others. His wastefulness consists primarily of an inability to make proper use of his advantages, and thus his personification of self-depletion differs from that of Raphaël de Valentin.

Each of the characters examined here offers in fact a different perspective on the effects of opposing forces which, as we have seen, may manifest themselves singly or in combination. When forces of release and self-sacrifice are positively personified, they are almost always found in conjunction with tendencies toward conservation. Madame de Mortsauf, for example, exemplifies both selfless dedication and repressive abstinence, whereas destiny holds Eugénie Grandet's giving nature in check. Vautrin's dedication, instead, is tempered by tireless vigilance and self-discipline. The latter's resourceful self-preservation contrasts both with Lucien's destructiveness and with Louis Lambert's deadly form of conservation.

If space allowed, it would be possible to continue almost indefinitely this inventory of Balzac's allegorical personifications: Foedora, Arabella, Nucingen, Madame d'Espard, and Nanon exemplify different aspects of conservation, and stand in opposition to such characters as Pauline, Coralie, Esther, David Séchard and Hulot.

Balzac's allegories are rooted in conflict and perpetuated, as we have seen, through dynamic transfers of energy. Energetic single-mindedness, and in some instances, outright monomania enables many of his characters to achieve allegorical status in *La Comédie humaine*. In reference to his own use of allegory, the German dramatist Hugo von Hofmannsthal has alluded to ". . . a certain timeless European mythology: names, concepts, figures, with which a higher meaning is bound up, personified forces of the moral or mythical order."[26] Balzac, in fact, constantly draws on this mythological patrimony while at the same time renewing it and increasing it. His realism is thus often ominously or majestically overshadowed by a mythic level of thought, causing his allegorical entities to ring truer and echo longer than they otherwise might.

University of Miami

NOTES

[1] C. S. Lewis, *The Allegory of Love* (rpt. Oxford: Oxford University Press, 1970), p. 60.

[2] Ernst Robert Curtius, *European Literature and the Latin Middle Ages*, trans. Willard R. Trask (rpt. New York: Harper Torchbooks, 1963), p. 205.

[3] *Ibid.*, p. 207.

[4] Honoré de Balzac, *Le Lys dans la vallée* (Paris: Gallimard, Le Livre de Poche, 1970), p. 96. All translations from the *Comédie humaine* are mine.

[5] *Ibid.*, p. 310.

[6] *Ibid.*, p. 321.

[7] Balzac, *Le Père Goriot* (Paris: Garnier-Flammarion, 1966), pp. 130–31.

[8] Balzac, *Splendeurs et misères des courtisanes* (Paris: Classiques Garnier, 1953), p. 511.

[9] Balzac, *La Cousine Bette* (Paris: Garnier, 1968), pp. 153–54.

[10] Diana Festa-McCormick, *Honoré de Balzac* (Boston: Twayne, 1979), p. 112.
[11] *La Cousine Bette, op. cit.*, p. 103.
[12] Balzac, *Eugénie Grandet* (Paris: Garnier Frères, 1965), p. 256.
[13] Balzac, *Louis Lambert* (Paris: Gallimard, Le Livre de Poche, 1968), p. 104.
[14] *Ibid.*, p. 128.
[15] *Ibid.*, p. 163.
[16] Balzac, *La Peau de chagrin* (Paris: Gallimard, Le Livre de Poche, 1966), pp. 233–34.
[17] *Splendeurs et misères des courtisanes, op. cit.*, p. 498.
[18] *La Peau de chagrin, op. cit.*, pp. 55–56.
[19] *Splendeurs, op. cit.*, p. 380.
[20] *Ibid.*, p. 479.
[21] *Ibid.*, p. 506.
[22] Balzac, *Illusions perdues* (Paris: Gallimard, Le Livre de Poche), p. 459.
[23] *Ibid.*, pp. 472–73.
[24] *Splendeurs, op. cit.*, p. 483.
[25] *Ibid.*, p. 491.
[26] Quoted by Curtius, *op. cit.*, p. 144.

HÉLÈNE VISENTIN

ORACLE ET ALLÉGORIE DANS *L'ANDROMÈDE* DE PIERRE CORNEILLE

[. . .] dans la chaleur que le ciel nous inspire,
Nos vers disent souvent plus qu'ils ne pensent dire;
Et ce feu qui sans nous pousse les plus heureux
Ne nous explique pas tout ce qu'il fait pour eux.
Remerciement à M. Le Cardinal Mazarin
(1644)[1]

En janvier 1650, le roi alors âgé d'une douzaine d'années assiste à la première représentation d'*Andromède*, tragédie à machines commandée par le Cardinal Mazarin à Corneille. Le mythe de Persée et Andromède, tiré des *Métamorphoses* d'Ovide, peut se résumer en quelques mots: Cassiope, femme du roi d'Ethiopie, est si vaine de sa beauté qu'elle ose la préférer à celle des Nymphes de la mer; ces dernières pour se venger invoquent Neptune qui libérera un monstre marin, dévastateur du royaume. Pour apaiser sa colère, l'oracle de Jupiter ordonne qu'Andromède, fille unique du roi et de la reine, soit dévorée par la bête. Attachée à un rocher et prête à être sacrifiée, Andromède est sauvée par le valeureux Persée qui plonge du ciel pour abattre le monstre.

Le faste des décors, les changements à vue ainsi que les "superbes Machines" du *Grand Magicien* Torelli ne peuvent qu'émerveiller l'enfant-roi à qui la pièce est adressée.

Toutefois, loin d'être un simple divertissement, *Andromède* exalte la puissance monarchique: dès le prologue, la figure du Soleil convie Melpomène, muse de la tragédie, à célébrer le roi:

Viens donc, viens avec moi faire le tour du Monde,
Qu'unissant ensemble nos voix,
Nous fassions résonner sur la Terre et sur l'Onde
Qu'il est le plus jeune et le plus grand des Rois. (vv. 69–72)[2]

"Louis est le plus jeune et le plus grand des Rois": tel sera le refrain du prologue.

Sans négliger le rôle de la flatterie qui assure au poète la protection royale, ne pourrait-on pas affirmer que le dessein de glorifier la monar-

M. Kronegger and A-T. Tymieniecka (eds.), Analecta Husserliana XLII, pp. 49–60.

chie en pleine Fronde est un acte politique, du moins une prise de position audacieuse par rapport au conflit qui dresse les princes contre la royauté?

Sans doute, mais ce n'est pas le lieu ici d'examiner les conditions politiques qui auraient pu motiver Corneille. Je tenterai plutôt de voir comment fonctionne le système allégorique dans l'*Andromède*. J'évoquerai rapidement la rhétorique de persuasion qui est mise en place au sein du texte dramatique, je dégagerai ensuite le sens moral de l'allégorie pour parvenir à montrer que dans cette pièce à machines, les exploits du héros Persée préfigurent ceux de Louis XIV. En ce sens, plus qu'une tragédie allégorique, *Andromède* serait un oracle qui fait du poète un prophète.

Dans cette tragédie à machines, Corneille introduit une rhétorique qui privilégie deux moments: celui du *probare* et celui du *movere*. Il s'agit précisément d'une rhétorique de persuasion qui opère à un double niveau: au niveau de la représentation par le merveilleux de la mise en scène et au niveau du texte par la figure de l'allégorie. Le merveilleux est sans doute la meilleure *captatio* capable de préparer les esprits à l'émotion et à l'agrément afin de mieux persuader. Christian Delmas écrit dans l'excellente introduction de son édition:

> . . . le merveilleux héroïque à la louange du roi se trouve dès *Andromède* détourné au profit d'un régime autoritaire en puissance, prêt à exploiter la crédulité des sujets, facilement éblouis par le faste du spectacle. Déjà en Italie le faste des représentations publiques est un "divertissement", au sens pascalien du terme, et un instrument d'asservissement.[3]

En effet, il est révélateur que sur la gravure de François Chauveau illustrant le décor de l'acte IV le palais royal d'Ethiopie est orné de la couronne et des armes de France.[4]

Pour convaincre son public de la puissance du Prince, Corneille établit habilement, au niveau de la nomination du héros, une correspondance héroïque entre le roi et Persée. Rarement appelé par son nom propre, il est désigné par des périphrases et des expressions qualificatives: "illustre inconnu", "parfait amant", "grand héros", "vaillant guerrier", ou encore par des formules qui ne vont pas sans rappeler le caractère divin que revêt la figure royale: Persée est tour à tour l'"enfant des Dieux", le "Seigneur" ou un "demi-Dieu". Sa situation est donnée d'avance comme celle du jeune roi, surnommé à sa naissance "Louis Dieudonné".

Dès les premiers vers de l'acte I, Cassiope s'adressera au héros en ces termes:

> Généreux inconnu, qui chez tous les Monarques
> Portez de vos vertus les éclatantes marques,
> Et dont l'aspect suffit à convaincre nos yeux
> Que vous sortez du sang, ou des Rois, ou des Dieux. (vv. 948–101)

Par ailleurs, les sentences qui parsèment le texte dramatique ont une grande force persuasive: elles expriment clairement la pensée de l'auteur qui veut convaincre de la loyauté du roi; des maximes telles que "le Roi dont tout dépend est homme de parole" (v. 1205), "le sang des Dieux vaut bien celui des Rois" (v. 250) et "[les lois] ne nous font part de leur sacré pouvoir / Que pour le mesurer aux règles du devoir." (vv. 298–299) sont dissimulées dans le discours des personnages.

Comme nous l'avons mentionné en introduction, dans *Andromède* les clefs nécessaires à la compréhension de l'allégorie sont livrées dès le prologue; Corneille guide le spectateur-lecteur dans sa recherche des sens cachés de l'oeuvre dramatique. A l'époque, nombreux sont les auteurs qui expliquent, dans les avant-propos, les préfaces ou les titres, le sens allégorique de leur ouvrage; dans un article sur la poétique de l'allégorie, Bernard Beugnot déduit que "le texte et ses gloses [. . .] constitue[nt] un tout aussi indissociable que le corps et l'âme de la devise".[5] Tel est aussi le fonctionnement de l'emblème politique.

Le mythe de Persée et Andromède a donné lieu à de nombreuses représentations emblématiques surtout sous les règnes d'Henri IV et de Louis XIII au moment des guerres de religions.[6] Signalons le placard de 1594 qui est paru peu de temps après l'entrée triomphale d'Henri IV dans la capitale, vainqueur de la Ligue et du pouvoir espagnol. Le sonnet qui accompagne la gravure est très explicite:

> FRANCE, comme Andromède à la mort fut offerte,
> Mal voulue des siens, et d'un peuple estranger,
> Son pays fut son mal, sa guerre et son danger,
> Où son bien devoit être, y demeuroit sa perte:

Le ciel fasché de veoir une injustice aperte,
Un Persée envoya à fin de la venger,
Un Persée françois qui la vint desgaigner
Des vagues de la mort, qui l'avoient ja couverte.

Le monstre qui gardoit entre ses dens sa mort,
Sentit combien le bras de Persée estoit fort,
Comme feit l'Espagnol de HENRY quatriesme.

FRANCE, sois luy fidelle, et ne te laisse plus
Attacher de doublons, et ne croy aux abus
De ceux qui ont rongner l'or de ton diadéme.[7]

Plus tard, l'*Entrée d'Arles* de 1622 décrit la victoire des troupes royales sur les protestants du Languedoc; des fragments du mythe ovidien sont représentés sur chacun des frontons des six arcs de triomphe[8]; Enfin, mentionnons le célèbre feu d'artifice qui glorifie Louis XIII en l'honneur de la victoire de la Rochelle (1628): le thème est celui de Persée délivrant Andromède. Voici le commentaire du sujet:

Votre Majesté [. . .] verra que cette pucelle attachée contre le rocher ne peut représenter autre chose que La Rochelle . . . Et ce vaillant Persée ne peut estre mieux figuré que par votre Majesté. Il estoit fils du plus grand des Dieux, et vous estes fils de celuy qui estoit le plus grand des Rois avant qu'il ne vous eust mis au monde; comme luy Vous avez seul mérité de triompher d'une Pucelle . . .[9]

Dans ces trois exemples (et on pourrait le vérifier pour d'autres événements historiques), on a recours au mythe ovidien pour établir à la fois une "continuité et une correspondance héroïques".[10] La représentation et le commentaire correspondent respectivement au corps et à l'âme de la devise.

Si sous Henri IV et Louis XIII le mythe marque l'accomplissement d'un exploit, sous Louis XIV il annonce l'exploit. En effet, Corneille conçoit *Andromède* pour le roi qui n'a pas encore remporté de victoires; mais dans le prologue, Melpomène les lui prédit et l'instruit, grâce aux modèles antiques, de ce qu'il sera plus tard. Elle annonce au Soleil à propos du "Monarque des Lys":

Que déjà je le vois, que déjà je lui plais,
Et que de ses vertus, et que de ses hauts faits,
Déjà dans ses pareils je lui trace une image.
Je lui montre Pompée, Alexandre, César,
Mais comme des Héros attachés à son char,
Et tout ce haut éclat où je les fais paraître
Lui peint plus qu'ils n'étaient et moins qu'il ne doit être.
(vv. 50–56)

Le modèle héroïque est présenté ici comme une image vraisemblable de Louis XIV: la muse offre à l'enfant-roi un miroir dans lequel il se voit agir.

C'est là qu'entre en jeu l'allégorie, figure de pensée que Fontanier définit comme

. . . une proposition à double sens, à sens littéral et à sens spirituel tout ensemble, par laquelle on présente une pensée sous l'image d'une autre pensée, propre à la rendre plus sensible et plus frappante que si elle était présentée directement et sans aucune espèce de voile.[11]

Andromède est donc une pièce à double niveau où coexistent le "littéral" et le "figuratif" et où s'élabore un système de correspondances: le héros mythologique a les traits du prince et le prince s'approprie l'image du héros.

En ce sens, l'allégorie a une fonction didactique et s'inscrit dans le mouvement de la pensée classique qui prône l'utilité de l'art: veiller à l'instruction du public en dissimulant une vérité profonde sous les travestissements de la fable. Corneille lui-même explique sa conception d'un art de la dissimulation:

L'œil se peut-il fixer sur la vérité nue?
Elle a trop de brillant pour arrêter la vue,
Et, telle qu'un éclair qui ne fait qu'éblouir,
Elle échappe aussitôt qu'on présume en jouir;
La Fable, qui la couvre, allume, presse, irrite,
L'ingénieuse ardeur d'en voir tout le mérite:
L'art d'en montrer le prix consiste à la cacher,
Et sa beauté redouble à se faire chercher.[12]

Si l'allégorie émeut, elle cherche aussi à prouver et dès lors elle tend à se développer, à se diffuser et à envahir toute la structure de la représentation. Le spectacle participe à la propagande royale: ce n'est que sous le voile allégorique que le discours politique s'énonce, comme si l'idéal monarchique ne pouvait être atteint qu'à travers le jeu infiniment complexe de la transposition et de la fabulation. Peu à peu, l'espace privé et public du roi va être envahi par le système allégorique; la monarchie se projette dans un imaginaire hérité des thèmes antiques "fixateurs d'éternité". La décoration des appartements et de la galerie de Versailles, conçue par Le Brun entre 1671 et 1687, en est peut-être le meilleur exemple d'autant plus qu'il illustre l'apogée du règne de Louis XIV; le "discours pictural", fondé sur des références mythologiques et historiques, donne à voir le roi et engendre une nouvelle vision du monde.[13]

De même dans *Andromède*, l'allégorie intervient dans le processus créateur. Elle fait partie de l'idée du poème, l'*inventio*: le mythe de Persée et Andromède se prête très bien à la glorification du jeune roi; et elle agence l'ensemble des idées poétiques, la *dispositio*: tout au long de l'action dramatique, Corneille doit établir des rapports constants entre une réalité dramatique et une dimension allégorique. L'allégorie n'est plus alors un simple *ornatus* rhétorique, mais un discours sur les valeurs morales. La tragédie de Corneille, par le processus d'identification du roi au héros mythologique, incarne les quatre vertus cardinales, généralement nommées depuis le moyen âge le "speculum principis".

La FORCE est symbolisée par l'épée et le bouclier de Persée qui délivre Andromède du monstre et combat Phinée, son rival. Ammon parle à son ami Phinée des armes de Persée en ces termes:

> Mais vous ne savez pas, Seigneur, que son épée
> De l'horrible Méduse a la tête coupée,
> Que sous son bouclier il la porte en tous lieux,
> Et que c'est fait de vous s'il en frappe vos yeux. (vv. 1306–1309)

Le sceptre et la couronne, attributs du roi Céphée qui incarne la figure du monarque accompli (sur les gravures, on le reconnaît facilement grâce à ces deux signes), sont les symboles de la JUSTICE:

> [. . .] heureux sont le Prince, heureux sont ses projets
> Quand il se fait justice ainsi qu'à ses sujets (vv. 306–307)

déclare le roi à ses proches.

Le bouclier en forme de miroir qui appartient à Persée est une allégorie de la PRUDENCE. "Le miroir n'est nécessaire que pour régler ses actions. L'Homme prudent examine ses défauts: ce qu'il ne peut faire sans la connaissance de soi-même" écrira Ripa dans son *Iconologia.*

Enfin, la vertu de la TEMPERANCE se retrouve dans le personnage de Persée: bien qu'il ait le consentement des dieux et du roi d'épouser Andromède, il ne s'impose pas et lui laisse le choix d'accepter son amour:

> De tout ce que j'ai fait, ne voyez que ma flamme,
> De tout ce qu'on vous dit, ne croyez que votre âme,
> Ne me répondez point, et consultez-la bien,
> Faites votre bonheur sans aucun soin du mien,
> Je lui voudrais du mal s'il retranchait du vôtre,
> S'il vous pouvait coûter un soupir pour quelque autre
> Et si quittant pour moi quelques destins meilleurs,
> Votre devoir laissait votre tendresse ailleurs. (vv. 1114–1121)

Ces vertus princières composent moins une image du parfait époux aux yeux d'Andromède qu'un portrait du prince idéal aux yeux de tous, et en particulier aux yeux de la reine, mère du roi, pour laquelle Vouet, en 1637, avait décoré les appartements du Château-Neuf de Saint-Germain-en Laye; les quatre vertus cardinales avaient été peintes dans des caissons décoratifs au plafond de sa chambre.

Par ailleurs, la conduite de Phinée (qui devait être le fiancé d'Andromède avant qu'elle ne soit désignée par l'oracle) exalte les vertus de Persée par antinomie. Phinée a tous les vices: il se révolte contre la volonté des dieux, devient lâche devant le danger, réagit égoïstement envers Andromède, tente d'assassiner Persée . . . Bref, il est l'autre monstre que Persée doit éloigner d'Andromède. Cassiope fera d'ailleurs remarquer à Phinée que, s'il avait voulu être à la hauteur, il aurait pu obtenir l'appui des dieux qu'a mérité Persée par son courage et sa promptitude:

> Le Ciel, qui mieux que nous connaît ce que nous sommes
> Mesure ses faveurs au mérite des hommes,
> Et d'un pareil secours vous auriez eu l'appui
> S'il eût pu voir en vous mêmes vertus qu'en lui.
> Ce sont grâces d'en haut rares et singulières,
> Qui n'en descendent point pour des âmes vulgaires,

Ou pour en mieux parler, la justice des Cieux
Garde ce privilège au digne sang des Dieux,
C'est par là que leur Roi vient d'avouer leur race. (vv. 1543–1551)

La pièce renferme donc tous les symboles des quatre vertus cardinales. Corneille a composé une allégorie du prince parfait: il offre ainsi au jeune roi un miroir à double facette: d'un côté, le modèle de Céphée, image du monarque absolu, et de l'autre, le modèle de Persée, image plus dynamique, qui enseigne à Louis XIV comment étonner par ses exploits s'il veut régner en en grand roi.

En fait, toute la pièce est discours: un discours moral qui s'énonce par les figures dramatiques et un autre discours, beaucoup plus subtil celui-là, qui s'exprime au niveau des oracles dans la tragédie; l'oracle est un discours dans le discours, c'est-à-dire un méta-discours.

Dans *Andromède*, nous pouvons relever deux oracles qui orientent l'action dramatique: ceux d'Ammon et de Vénus. Mais dans la mesure où Persée préfigure le roi-soleil, j'émets l'hypothèse que la tragédie à machines de Corneille est un troisième oracle qui s'adresse directement à l'enfant-roi pour qui la pièce est d'abord conçue. En ce sens, *Andromède* s'inscrit dans la succession des horoscopes et des songes qui ont jalonné toute l'enfance de Louis XIV, dans l'intention de lui prédire un avenir hors du commun et de lui façonner une image royale exceptionnelle.

Le texte littéraire du XVIIe siècle, tout comme l'ensemble des arts figuratifs, est un discours qu'il nous faut décoder. A ce propos, citons un texte de Baudoin de 1627 qui nous renseigne sur "le profit qu'apporte la connaissance des Fables':

. . . Quant à moi, j'estime que l'invention des Fables est comme un très doux assaisonnement de la vie humaine et qu'elles ne soulagent pas de peu les afflictions qui nous surviennent en ce monde; et je crois que tel fut le dessein des Anciens en la composition d'icelles. Car elles nous fournissent avec un singulier plaisir des enseignements pour régler notre vie, auxquels, n'était le plaisir des Fables, nous tournerions bientôt le dos. Ceux qui n'éplucheront pas de près le sens moral des Fables, et qui ne s'attachant par manière de dire à la première écorce, ne penseront pas qu'il y ait rien de plus divin caché là-dessous, ne pourront en recevoir cette utilité. Car ceux-ci se séant près du feu, comme font les enfants en hiver, se repaissent de contes de vieilles et de je ne sais quelles fables des poètes, ne se souciant au reste du principal sens, et de la plus profitable doctrine qu'il en fait extraire.[14]

C'est ainsi qu'à partir des trois oracles d'*Andromède*, nous pouvons mettre en évidence les différents nixeaux de sens de l'allégorie classique.

L'oracle d'Ammon commande le niveau littéral:

> Pour apaiser Neptune, exposez tous les mois
> Au Monstre qui le venge une fille à son choix,
> Jusqu'à ce que le calme à l'orage succède:
> Le sort vous montrera
> Celle qu'il agréera;
> Différez cependant les noces d'Andromède. (vv. 176–180)

Ces vers exposent l'intrigue principale de la fable dramatique: c'est la figure du monstre qui est au premier plan.

Vénus descendra au milieu des mortels pour annoncer le second oracle:

> Andromède ce soir aura l'illustre époux
> Qui seul est digne d'elle et dont elle est digne,
> Préparez son Hymen, où pour faveur insigne
> Les Dieux ont résolu de se joindre avec vous. (vv. 358–361)

Le rôle de cet oracle est de provoquer le choix d'un digne époux. L'action dramatique substituera à Phinée un vrai héros, Persée. On aborde ici le sens moral: c'est dans la peinture des vices et des vertus que se trouve "l'utilité du poème dramatique". Objet de la quête amoureuse, Andromède représente l'enjeu de l'oracle.

Enfin, l'oracle de Corneille commande le niveau allégorique de la pièce: Persée est l'allégorie du roi. L'oracle propose à l'enfant royal un programme à accomplir pour devenir le plus grand des rois. La référence aux modèles antiques est la forme privilégiée du "discours héroïque": Corneille emprunte son sujet à une réalité déjà ancienne afin de persuader son public que Louis XIV ne tardera pas à être supérieur aux héros de l'Antiquité; le Soleil dira à Melpomène à propos du jeune roi:

> Il en effacera les plus glorieux noms,
> Dès qu'il pourra lui-même animer son armée,
> Et tout ce que d'eux tous a dit la Renommée
> Te fera voir en lui le plus grand des Bourbons. (vv. 57–60)

L'allégorie impose donc une "manière de lire" qui permet de déjouer les pièges de la fable et d'accéder à la vérité profonde de l'œuvre.

Remarquous que chacun des oracles privilégie une figure dans un ordre qui respecte le récit dramatique: l'oracle d'Ammon privilégie le Monstre, l'oracle de Vénus, Andromède et l'oracle de Corneille, Persée. La tragédie à machines de Corneille s'offre comme un tableau qui s'inscrit dans la tradition picturale du mythe où l'on voit la disposition linéaire ou triangulaire de ces trois figures au moment où Persée plonge du ciel pour abattre le monstre.[15] En fait, ces trois figures correspondent moins à des moments spectaculaires qu'à des moments de discours: l'oracle d'Ammon, l'oracle de Vénus et bien sûr l'oracle de Corneille, ultime discours de cette pièce à machines.

Ainsi, l'imaginaire monarchique représenté dans *Andromède*, loin de réduire le modèle mythologique, le transpose et lui donne une dimension nouvelle: Louis XIV, sous le masque de Persée, se révèle être l'allégorie de roi parfait. La représentation par le discours dramatique persuade de la puissance du prince et lui garantit universalité et éternité.

C'est à partir des années 1660–1663 que tous les talents artistiques serviront la propagande louis-quatorzienne. Les premières victoires militaires en même temps qu'elles perpétuent l'héroïsation du roi, confirment toutes les prédictions de son enfance.

En guise de conclusion, je donnerai le dernier mot à Corneille qui s'adresse au roi par ces vers:

> Mon génie au théâtre a voulu m'attacher;
> Il en a fait mon fort, je dois m'y retrancher;
> Partout ailleurs je rampe, et ne suis plus moi-même;
> Mais là j'ai quelque nom, là quelquefois on m'aime;
> Là ce même génie ose de temps en temps
> Tracer de ton portrait quelques traits éclatants.
> Par eux de l'*Andromède* il sut ouvrir la scène;
> On y vit le Soleil instruire Melpomène,
> Et lui dire qu'un jour Alexandre et César
> Sembleraient des vaincus attachés à ton char:
> Ton front le promettait, et tes premiers miracles
> Ont rempli hautement la foi de mes oracles.[16]

Université de Montréal

NOTES

[1] *Remerciement à Monseigneur l'éminentissime Cardinal Mazarin* (1644) in: *Oeuvres complètes* de Pierre Corneille, éd. par G. Couton, Paris, Gallimard, "Bibl. de la Pléiade", 1987, Tome I, p. 1063, v. 33–36.
[2] Toutes les références au texte de Corneille renvoient à l'édition établie par Christian Delmas: *Andromède*, Paris, Didier, Société des Textes Français Modernes, 1974, 217 p.
[3] Cf. Introduction d'*Andromède*, éd. citée, p. CI.
[4] Illustrateur reconnu à l'époque, François Chauveau grave les planches des six décors ainsi que le frontispice de l'*Andromède*. C'est lui d'ailleurs qui exécutera 16 des 23 gravures qui illustrent l'édition de 1660 des œuvres de Pierre Corneille.
[5] "Pour une poétique de l'allégorie classique", *Critique et création littéraire au XVIIe siècle*, Paris, CNRS, 1977, p. 415.
[6] Le mythe ovidien est très flexible et permet d'illustrer l'actualité du moment: Andromède repésente généralement la France persécutée par le monstre de l'hérésie et Persée, le bon Prince qui défend son royaume.
[7] Cité par Françoise Bardon in: *Le portrait mythologique à la cour de France sous Henri IV et Louis XIV*, Paris, 1974, pp. 133–135.
[8] Wagner, Marie-France, "L'icône politique dans l'entrée d'Arles de 1622", *Paragraphes*, n° 26, 1900–1991, pp. 51–61.
[9] Bardon, Françoise, *op. cit.*, p. 281.
[10] J'emprunte l'expression à Françoise Bardon.
[11] Fontanier, *Les figures du discours*, Paris, Flammarion, 1977, p. 114.
[12] "Défense des Fables dans la Poésie" in *Oeuvres complètes*, éd. citée, Tome III, p. 733, v. 41–48.
[13] À ce sujet, voir l'article de Gérard Sabatier, "Le parti figuratif dans les appartements, l'escalier et la galerie de Versailles", *XVIIe siècle*, oct.–déc. 1988, n° 161, pp. 401–427.
[14] *Mythologie ou explication des fables*, Paris, 1627 (BN. J. 1943).
[15] Depuis le milieu du XVIe siècle surtout, de nombreux peintres ont représenté le mythe de Persée et Andromède; signalons, parmi les plus célèbres, Titien (1554–1556), Véronèse (1584), Mignard (1679) et Nicolas Bertin (1715).
[16] "Remerciement au Roi" (1663) in *Oeuvres complètes*, éd. citée, Tome III, p. 453, vv. 31–42.

REFERENCES

Bardon, Françoise, *Le portrait mythologique à la cour de France sous Henri IV et Louis XIII, mythologie et politique* (Paris: Editions A. et J. Picard, 1974), 326 p.

Beugnot, Bernard, "Pour une poétique de l'allégorie classique", in: *Critique et création littéraires en France au XVIIe siècle*, Paris, CNRS, 1977, pp. 409–419.

Charbonneau, Frédéric, "La face cachée de l'amante: lecture rhétorique de l'*Andromède* de Pierre Corneille", in: *Paragraphes* **6**, 1990–91, pp. 87–99.

Corneille, Pierre, *Andromède*, texte édité par Christian Delmas et enrichi de six illustrations (Paris: Didier, "STFM"), 1974, 217 p.

Corneille, Pierre, *Oeuvres complètes*, éd. par G. Couton (Paris: Gallimard, "Bibl. de la Pléiade", 3 vol., 1983–1987).

Couton, Georges, *Corneille et la fronde: théâtre et politique il y a trois siècle (*Clermont-Ferrand, Imprimerie de Bussac, 1951), 112 p.

Couton, Georges, "Réapprendre à lire: deux des languages de l'allégorie au XVIIe siècle" in: *Cahiers de l'Association Internationale des Etudes Françaises*, **28**, mai 1976, pp. 81–101.

Ferrier, Nicole, "Comment un roi devient héros: Louis XIV dans la littérature française de son temps", *Trois figures de l'imaginaire littéraire: les odyssées, l'héroïsation de personnages historiques et le savant*, Actes du XVIIe congrès (Nice 1981) de la Société Française de Littérature Générale et Comparée, recueillis et publiés par Edouard Gaède, Paris, Les Belles Lettres, 1982, pp. 167–181.

Fontanier, *Les figures du discours* (Paris, Flammarion, 1977), 505 p.

Ovide, *Les Métamorphoses*, éd. par Joseph Chamonard (Paris, Garnier, 1966) 503 p.

Sabatier, Gérard, "Le parti figuratif dans les appartements, l'escalier et la galerie de Versailles", in: *XVIIe siècle*, oct-dèc 1988, **161**, pp. 401–427.

Wagner, Marie-France, "L'icône politique dans l'Entrée d'Arles de 1622", in: *Paragraphes* **6**, 1990–1991, pp. 51–61.

Le peintre, le roi, le héros. L'Andromède de Pierre Mignard, présenté par Jean-Claude Boyer, Paris, éd. de la Réunion des musées nationaux, 1989, 107 p.

JOAN B. WILLIAMSON

ALLEGORY THEN AND NOW: THE PHYSICIAN AND DISEASE

As Norman Cousins has remarked:

> It should be no surprise that almost every novelist or dramatist of any consequence – from Aeschylus to Walker Percy – has had something to say about doctors. The writer deals with the universals of human experience – the struggle not just to stay alive but to get the most out of life is at the epicenter of those universals. To the writer, the physician is not just a prescriber of medicaments but a symbol of all that is transferable from one human to another short of immortality.[1]

Given this literally vital importance of disease and the healing figure, the study of the ways in which they are treated allegorically provides a wealth of material to illustrate differences between medieval and modern allegory.

The physician today has acquired such exalted status that he or she has virtually become a superhuman figure. Consider only the case of TV's physician Marcus Welby, whose aura of authority is so great that the actor who portrays him sells pharmaceuticals in TV advertisements. Even the malpractice lawsuits, encouraged by the fees physicians command, are themselves reflections of this attitude, since their point of departure is that the deliverer of care has no right to fail. Astronomical fees, together with the physician's desire to inspire confidence, proclaim that he or she is omnipotent. Legal fees are equally high, but here the public seems more conditioned to dubious outcomes. While such TV programs as "LA Law" portray lawyers in a highly favorable light, the public seems ready to define lawyers as crooks and con artists who are experts at relieving the client of his or her money. The old image of the lawyer as the monkey who adjucates the two goats' request for equal division of a piece of cheese by nibbling away at each half – to equalize them – until there is nothing left, still prevails in the minds of many.

The negative image of the bad physician is also old, for we find in the Bible: "But ye are all forgers of lies, ye are all physicians of no value" (Jb 13:4). Such negative images flourished in the medieval world, one such pejorative image depicting the physician as money-oriented, with disparaging remarks recorded about denial of treatment until payment in full had been made. In witness of this attitude we have Chaucer's

M. Kronegger and A-T. Tymieniecka (eds.), Analecta Husserliana XLII, pp. 61–82.

immortalization of the physician's love of money in the General Prologue to the *Canterbury Tales*, where he recounts: "He kept that he wan in pestilence./ For gold in phisik is a cordial,/ Therefore he lovede gold in special" (vv. 442–44).[2] Philippe de Mézières, commenting on justice that is only so in appearance, compares this kind of justice to the situation of skinned lambs handed over to be judged by ravenous wolves, and to the care a poor man receives if he does not have the wherewithall to pay his physician, as if this last situation were so well-established as to provide an obvious comparison:

Mais pour moustrer es dictes requestes aucune apparance de justice, on renvoioyt les aigneaux eschorchez au jugement des loups ravissables, c'est assavoir les pauvres gens susdiz aux esleuz et chevetaines des aides; desquelx les pauvres gens se trouveoient aussi reconfortez comme le grief malade du phisicien qui n'a de quoy lui satisfaire ne le payer de son salaire.[3]

Philippe further excoriates physicians, this time those who treat princes and great ladies, to whom great ills can befall if their physician is too greedy.[4] While respecting good physicians, he warns that the King should be sure his physician is well qualified, discreet, of good reputation, and neither too greedy nor ambitious so that he will not hold his patients hostage to great sums of money.[5] Eustache Deschamps, in his *Lal des Douze Estas du monde*, enumerating the duties of the twelve estates, notes that all are remiss, including the physician: "Medicin fault qui assouaige/ Les maulx des gens, sanz nul oultraige,/ Qui soit expers, de bonne vie" (vv. 85–87).[6] Thus does medieval literature eloquently record the negative image of the physician.

In contrast, a more positive image dominates the modern imagination. The reverence for the physician, viewed as the restorer of life, stems in part from the desire for the all-protecting parent who drives away the problems of one's infancy combined with the often dire nature of the problems that make one seek the services of the physician.

However, this attitude also reflects a survival of the more deeply held allegorical view of the physician as serving the divine, in an imitation of Christ in his practice of the art of healing. The very origin of medicine was believed to be divine: it was initially Apollo who had the art of curing, which he then passed on to his human son Aesculapius. But, after the latter was killed by a thunderbolt, the art of healing was forbidden to mortals and was not practised for some five hundred years.[7] Pythagoras (580–498 B.C.) founded a system of medicine and two of his pupils, Alcmaeon and Empedocles, who were physicians, presented the theory of the four elements, earth, water, fire, and air. Hippocrates (460–355 B.C.), who claimed to be a descendant of Aesculapius, added to the four elements the four humors: blood, phlegm, yellow and black bile, and the four conditions of hot, cold, moist, and dry.[8] But the connection of medicine with religious practice also continued. In Rome, medicine was left to foreigners, first Greeks, then Jews, then Gentiles who had joined the liberal Jewish communities in Rome, and who, when they became Paul's converts to Christianity, became the Christian medical care-givers of the Empire.[9]

The figure of Christ as healer thus has a background of association of medicine with the divine, but it also draws strong authentication from the Bible. In the Old Testament we have the incident where God made Moses' hand leprous and healthy by turns (Ex 4:6–7). In the New Testament it is Christ who is portrayed as the healer. For example, in Matthew there is the healing of the two blind men (MT 9:27–31). In Mark there is the curing of the leper (Mk 1:40–45), the curing of the deaf and dumb man (Mk 7:31–37), and the healing of the blind man (Mk 8:22–26). In Luke, the patron of physicians because of the number of Christ's healing miracles that he records, we have the woman's infirmity cured (Lk 13:11–17), a case of dropsy healed (Lk 14:1–6), the ten lepers healed (Lk 17:11–19), the blind man begging by the wayside whose sight is restored (Lk 18:35–43), and the restoration of Malchus' ear (Lk 22:50–51). In John we find the nobleman's son cured of fever (Jn 4:46–54), and the man born blind cured at Jerusalem (Jn 9:1–7). Christ is also shown as healer in the Epistles, as in James 5:13–15. But not only did Christ heal, he raised the dead to life, restoring life to Lazarus who had been buried in the grave for three days (Jn 11:38–44). Indeed Christ rightly called himself physician when he cited the proverb: "Physician, heal thyself" (Lk 4:23).

The figure of Christ as supreme Physician was established in the Sacred Scriptures. But medical terminology concerning the ills of the mind and soul had been used throughout antiquity.[10] Well-known to the Church Fathers, e.g., St. Augustine writes: "I hide not my wounds; Thou art the Physician, I the sick" (*Confessiones* 10.28), this usage continued into the Middle Ages.[11] The idea of Christ as physician of the soul was taken up in allegorical vein by the two first great Christian mystics, Hugh of St.-Victor and St. Bernard of Clairvaux. Hugh of St.-Victor wrote his *De medicina animae*, elaborating a complex spiritual framework couched in medical terminology and based on the concept of the four humors;[12] while St. Bernard frequently resorted to the image of the physician to present the loving Christ, e.g., "the heavenly physician came with speed to her aid, because 'his word runs swiftly.' Perhaps you think the Word of God is not a medicine? Surely it is, a medicine strong and pungent, testing the mind and the heart" (*Sermon on the Song of Songs* 3.1–2).[13] Thus was launched the allegorical pattern of Christ as healer of souls and the allegorical equation of sin and disease to which we will return later.

The Acts of the Apostles also authorize belief in the transferral of Christ's power to his Apostles, as when Peter healed a man born lame (Acts 3:1–8), when he cured Aeneas of his palsy (Acts 9:33–34), and when he raised Tabitha from the dead (Acts 9:36–41). We have the tradition of the physician following in Christ's footsteps and on occasion performing miracles, as in the case of Cosmas and Damian, the third century Arab brother-physicians who refused remuneration for their services and whose most celebrated case was what was perhaps the first recorded transplant, where they cut off the gangrenous thigh of a white patient in Rome and attached in its place the thigh from a newly-buried black corpse. When the patient woke up he found that he had a healthy but black thigh and that his rotted leg was attached to the corpse.[14]

This transferral of Christ's powers to his followers is at the heart of the Church's canonization process, in which miraculous cures achieved through contact with the corpse or articles having belonged to the dead person attest to that person's sanctity. We see Philippe de Mézières attach to his life of Saint Peter Thomas, the papal legate to Cyprus, a list of miracles wrought by the body of the saint. Graphic details of the physical state of the sufferers are provided as proof of the miraculous nature of the cure, e.g., the case of Brother John, who, suffering from quinsy, had such a sore throat that for nine days he could neither eat nor drink nor sleep, but who was cured when the right hand of the papal legate was placed on his throat.[15] Popular tradition, perpetuating the idea of transfer of divine healing powers to privileged human beings, credited some saints with the ability to miraculously heal specific diseases associated with them, e.g., St. Roch protected against and sometimes cured plague, while St. Hubert cured rabies from wolf and dog bites.

The transferral of Christ's power to his followers, making of them agents of the divine power, also extends to kings, who were believed to be his representatives on earth. From this the belief in royal thaumaturgy developed during the eleventh and twelfth centuries in England

and France, where it persisted for several centuries. Marc Bloch records that the disease scrofula,[16] which could be cured by the royal touch, was endemic in the Middle Ages.[17] He traces documentary evidence of the French rite to Louis VI (who reigned from 1108 to 1137) and his father Philip I (1060–1108), through the attestation of Guibert of Nogent;[18] and of the English rite to Henry II (who died in 1189) through a letter by Peter of Blois.[19] Although we have the denial of William of Malmesbury, who maintained that Edward the Confessor's miraculous healing power came from his sanctity and not his royal blood,[20] Shakespeare's report on Edward's healing touch (Macbeth, IV, iii) upholds the tradition that the miraculous royal power was already exercised by Edward and then by Henry I.

Philippe de Mézières offers an example of this healing power of the king when he reminds Charles VI of France that this kingship confers on him the power to heal by virtue of his holy annointment, so that if he merely touches with his hand a patient sick with scrofula, which is like a leprosy, the patient will be healed without any medication in what is an evident miracle:

> Que pour touchier tant seulement de ta main sacree a la maladie qui s'appelle escroelles, laquelle en son membre et en sa partie vault bien une mesellerie, le pacient, sans medicine, en certain temps recoit plaine sancte, qui est une miracle evident, par la vertu de la sainte unction, a la loenge de la sainte Trinite.[21]

But Philippe also developed the figure of the king healing by virtue of his royal heritage allegorically, to make of Charles VI the healer of the economic and social ills of his country. Charles VI is admonished by our author to be a sacred physician: "comme bon phisicien par le commandement de Dieu tu as emprins la cure."[22] As such he must administer harsh and bitter medicines to some and soothing and fortifying drugs to others, which are of course economic and political measures.[23]

Of the various kinds of diseases, not all were or are treated in allegorical fashion. In considering modern allegorical treatments of disease we are fortunate in having Susan Sontag's two books on the subject: *Illness as Metaphor* and *AIDS and Its Metaphors*.[24] From these we learn that in the nineteenth and twentieth centuries, only tuberculosis, syphilis, cancer and now AIDS have been treated metaphorically.

We would add that so far AIDS has not generated much literature, but it has nonetheless been subjected to an allegorical process. Susan

Sontag has pointed out how in popular parlance AIDS is believed to be caused by perverse sex or drugs, ignoring African heterosexual spread;[25] and how infectious diseases to which sexual fault is attached inspire fears of easy contagion and bizarre fantasies of non-venereal means of transmission.[26] Recent newspaper reports bear out her argument. *The New York Times*, November 18, 1991, reports that in Black churches AIDS for many has overtones of sexuality and sinful behavior,[27] and Sunday January 19, 1992, cites a report by the National AIDS Program office showing that courts are guided more by stereotypes and misperceptions than by scientific evidence, as evidenced by the fact that persons with AIDS have harsh penalties meted out to them for biting and spitting at other people, even though such conduct poses minimal risk of transmission.[28]

The nineteenth century romanticized disease. Sontag has pointed out that there was the idea that the brain lesions of neurosyphilis inspired original thought. She reminds us that for Thomas Mann:

> whose fiction is a store-house of early 20th C. disease myths, syphilis is a muse central to his *Doctor Faustus* with its protagonist a great composer whose voluntarily contracted syphilis won from the Devil the promise that the infection would be limited to the central nervous system and would confer twenty-four years of incandescent creativity.[29]

In her earlier book, *Illness as Metaphor*, Sontag has described the romanticization of tuberculosis. She describes the visible symptoms of TB, the red blood on the handkerchief, its spells of euphoria, increased appetite, exacerbated sexual desire, rosy cheeks – a sign of fever though taken by some for a sign of health.[30] She sees tuberculosis as a disease of time, speeding up life (for it gallops), highlighting and spiritualizing it.[31] Mimi in *La Bohème* and Marguerite Gautier in *La Dame aux camélias*, Smike in *Nicholas Nickleby*, all die "the painless, refined, unfrightened, beatific deaths" of the tubercular patient,[32] for tuberculosis takes on qualities assigned to the lungs, which are part of the upper, spiritualized body and hence "a disease of the lungs is metaphorically a disease of the soul."[33] Thus tuberculosis can be used, totally unrealistically, to aestheticize death.[34]

On the other hand, Sontag sees cancer as a disease of the body that can strike anywhere, often the parts of the body one is embarrassed to name.[35] She defines cancer metaphorically as a pathology of space, spreading and mutilating parts of the body.[36] A cancer death is considered painful, a hard death, where "the dying cancer patient is portrayed as robbed of all capacities of self-transcendence, humiliated by fear and agony."[37] Cancer has also been metaphorized, she claims, pointing to psychological theories that make cancer patients responsible for their own disease.[38] A cancer, according to the metaphoric use of disease, is something evil, seen as one's fault, due to unsafe behavior, smoking, indulging oneself, due to weakness.[39] She asserts that the diseases that produce hard deaths through a dehumanizing process are the ones fraught with the most terror,[40] and, acquiring meanings that give form to our deepest fears, inflict stigma.[41] Sontag's point is that this way of looking at disease leads to even more suffering, and so she seeks to separate metaphors from disease. This is a laudable aim and a correct evaluation of certain attitudes to disease that do make it worse for the sick.

Nonetheless, the allegorical process is innate in the human mind, and the allegorical treatment of disease pervades literature, often to positive effect. The Middle Ages did not romanticize disease as did the nineteenth century, but it did present it allegorically, using disease to represent something else, and using other images to portray disease.

In examining the medieval allegorical treatment of leprosy, let us consider how St. Louis likened leprosy of the body to mortal sin on the soul, preferring the former to the latter since, as he tells us, mortal sin is the leprosy that remains on the soul for as long as God exists:

> Et bien est voir que quant l'omme meurt, il est gueri de la meselerie du cors. Mes quant l'omme qui a fait le pechié mortel meurt, il ne sait pas ne n'est certeins que il ait eu [en sa vie] repentance que Dieu li ait pardonei, par quoy grant poour doit avoir quo celle mezelerie li dure tant comme Diex yert en paradis *(Vie de Saint Louis* 28).[42]

The image fits exactly because the identification of physical leprosy is considered as a legal death, resulting in banishment, loss of property, possessions and the right to the exercise of one's profession.[43] Mortal sin in like manner separates the soul from communion with God in whom all good things reside.

However, I believe Sontag is oversimplistic in equating modern feelings connotated with the term leprous with medieval attitudes, when she says that feelings of evil are unnecessarily projected onto the diseases.[44] The disease inspired fear and horror precisely because it was considered a deadly disease.[45] In actual fact it almost certainly was a deadly disease then because such trauma as the involvement of the lungs and infections resulting from leprous mutilations did not have the benefit of today's advanced treatments. It was also almost certainly more transmissible in the Middle Ages, with its crowded and unsanitary living conditions, than now.

The leper was consequently shunned in the Middle Ages to avoid contagion, but not so totally as we might think, for lepers were not totally banned from circulating in towns. As Saul Nathaniel Brody has pointed out, sequestration laws were frequently unenforceable, or at least unenforced.[46] In Thomas's *Roman de Tristan* when Tristan seeks to secretly visit Iseut, he disguises himself to look like a leper and thus is able to approach the Queen (vv. 502–560).[47]

However, while moralists inevitably used leprosy as a symbol of sin because of Christ's healing of lepers in the Gospels, there is evidence to suggest that the medieval leper was not always necessarily deemed morally reprehensible by association with his disease. The lepers in *Le Roman de Tristan* by Béroul are a case in point. The afflicted Yvein is seen with one hundred companions, all with crutches (puioz, v. 1160) and sticks, all hunchbacked and deformed, all bearing clappers (vv. 1154–94). The gist of Yvein's speech to King Mark when he sees that Iseut is to be burned at the stake is that the King should give Iseut to the lepers, which would be a greater punishment. Explaining that the lepers have such ardor that no woman can tolerate sex with them, Yvein points to their horrifying bodies with the clothes stuck to their skins; and contrasts her prior gentle living, in a great house with good food and drink with the lepers' miserable lives and badly furnished hovels:

Sire, en nos a si grant ardor
Soz ciel n'a dame qui un jor
Peüst soufrir nostre convers:
Li drap nos sont au cors aers.
. . .
Se la donez a nos meseaus,
Qant el verra nos bas bordeaus
Et eslira l'escouellier
Et l'estovra a nos couchier
(Sire, en leu de tes beaus mengiers
Avra de pieces, de quartiers
Que l'en nos envoi'a ces hus) (vv. 11594-1209).[48]

It has been suggested by some that medieval lepers were over-sexed, perhaps because such scholars have equated medieval leprosy with a virulent form of syphilis, a venerally spread disease, with skin lesions like yaws, an African disease, bacteriologically indistinguishable from syphilis since it has the same causative organism, but which is spread mainly by non-venereal contact.[49] There was, in all events, a long ecclesiastical tradition which made physical diseases emblems of moral defects,[50] where leprosy was traditionally portrayed as punishment for wanton and lustful living.[51] Because of the uniformity produced by one Church, this imagery prevailed and medieval leprosy has "indelibly marked attitudes, language, and literature."[52] On the other hand, while medieval writers warn that lepers burn with desire for sexual intercourse, the really sick leper lost his sexual desire along with his health.[53] Lepers were segregated and the image of the sexually-charged leper is more probably simply a reflection of the condition of men deprived of commerce with women for a long time, which is identical to that of sailors returning from a long ship's journey, during which they often resorted to homosexual practices.[54]

The meaning of The Fall is that through sin death, and with it disease, came into the world, and therefore diseases naturally lend themselves to the allegorical treatment of sin. But we must not attribute to the medieval mind the literal linking of particular sins with specific diseases in an exclusively causal relationship. Christ made clear that while calamities may invite people to repent, sin is not their immediate cause (Lk 13:1–5). Specifically, as Jerre Collins of the University of Wisconsin reminded me, Christ unequivocally disassociated disease from sin when he stated that neither the sufferer's sins nor those of his father had caused a man's illness: "As he went along, he saw a man who had been blind from birth. His disciples asked him, "Rabbi, who sinned, this man or his parents, for him to have been born blind?' 'Neither he nor his parents sinned,' Jesus answered 'he was born blind so that the works of God might be displayed in him' (Jn 9:1–3). Medieval moralists could not have been unaware of this teaching, and consequently, when they presented sins in terms of leprosy and other diseases as a way of concretizing the horror of evil, we must not conclude that they were necessarily intimating that all lepers had fallen victim to the disease because of their lustful ways.

Literature supports my premise. For example, in Béroul there is no sense that the lepers are guilty of anything. They are just sick and this is their life, in the same way that the plague in Camus' *La Peste*, as Sontag herself has pointed out, is not an allegory of Nazi occupation, "it simply is . . . it gives life its seriousness."[55] What this incident shows is how poorly the lepers lived – their misery was real because they had neither property nor wealth, and had to live as outcasts from society. If Béroul's lepers are sexually frustrated, it is because of their segregation. The statement that Tristan will not hurt them bears out their innocence: "Tristran n'en vost rien atochier/ Ne entester ne laidengier" (vv. 1257–58). An innocence that Béroul stresses with his comment that while some story-tellers say they drowned Yuain, he, Béroul, holds the story better in his memory and says that Tristan was too worthy and too courteous to kill such people:

> Li conteor dïent qu'Yvain
> Firent nïer, qui sont vilain;
> N'en sevent mie bien l'estoire,
> Berox l'a mex en sen memoire,
> Trop ert Tristran pruez et cortois
> A ocirre gent de tes lois (vv. 1263–70).

We have the testimony of real-life lepers concerning attitudes towards them and their disease, which also shows that the medieval mind did not always equate the sufferer with the disease, nor make him morally responsible for its occurrence. These witnesses, because they were poets, have recorded their impressions of the disease and its effect on their lives. Jehan Bodel and Baude Fastoul took their leave of the world in poems when they were forced into exile because they had contracted leprosy.[56] Jehan Bolel composed his poem probably some time after 1200, dying some ten years later at the end of February or early March of 1210.[57] Baude Fastoul wrote in the second half of the 13th century and is believed to have died in 1270 or 1272.[58] Ironically Jehan Bodel says goodbye to Baude Fastoul: "Bauduïn Fastoul, or m'enplaide" (str. 28, v. 325), who, struck with the same inexorable disease, will later write his own leave-taking poem, referring in his turn to his senior fellow-poet Jehan Bodel, whose fief he sadly comments he must receive ("Ke je doi recevoir le fief/ Ki vient de par Jehan Bodel" (str. 19, vv. 227–28).

Since each poet refers to the other, we know that both came from the confrérie of poets in Arras. Thus we have reason to believe that Susan Sontag is wrong when she says: "Leprosy, very rarely fatal, and extremely difficult to transmit was not much more so when at its greatest epidemic strength" between ca: 1050–1350.[59] Here we have Jehan Bodel and Baude

Fastoul, both from the confrérie in Arras and both with leprosy, with the first dying within ten years of the acknowledgement of his disease at a time when it was not in as advanced a stage as was Baude Fastoul's. All the evidence points to medieval leprosy being a different form than that seen today, and it may well have been more contagious. Further, with people living in close quarters, with no running water for bathing and no antiseptic sewage system, any disease would spread more readily once in place. Leprosy was seen as an incurable, deadly disease, and feared as highly contagious. In this light, denial of a doomed person's rights in advance of his actual death to protect the healthy must have seemed a reasonable stance.

Both of our poets insist on leprosy's incurability. Jehan Bodel notes that he can only get worse: "Enferté et poison et plaie/ M'a donné pour le cors despire./ Or m'estouvra gaitier le pire" (str. 7, vv. 78–80). And while he acknowledges the valiant efforts of his physician, they are in vain (str. 26, vv. 301–12). Indeed all the physicians of Salerno together will not reduce his suffering: "Mais Diex m'a joué d'un sophisme/ Que tuit le mire de Salerne/ N'abaisseroient ceste lime" (str. 17, vv. 200–202).

Similarly, Baude Fastoul sees his disease as incurable: "Mes maus ki est tournés a plane,/ Dont cascuns dist que nus ne sane," (str. 15, vv. 169–170). This second poet, whose leprosy was considerably more advanced, perceives an openly bleak prognosis: his leg is eaten away: "M'ont une gambe si destruite" (str. 16, v. 190); his face is invaded: "Puis que ma face est entamee" (str. 20, v. 230). He sees himself as getting worse: "Car je vois en empirement;" (str. 12, v. 138); and pictures himself as ultimately only a healthy heart in an eaten-away body: "Quant je n'arai ne pié ne main,/ Bouce ne nés, fors le cuer sain,/ Dont dira li ame Deus vaille!" (str. 7, vv. 82–84).

Both these poets see their disease as sent by God so that they may expiate their sins and save their souls: Jehan Bodel: "Diex, qui tous biens acoustumas,/ Qui de ta verge batu m'as" (str. 30, vv. 352–53); and Baude Fastoul: "Pour mes griés peciés espenir" (str. 2, v. 21): and "Que Dix en me carnalité/ Avoit mis par humilité/ Un mal dont nus ne se renvoise" (str. 17. vv. 202–204). Bowing to the divine will, Baude Fastoul, as did Jehan Bodel, recognizes that his leprosy is sent by God as his just desserts: "Mais Dix m'envoie par desserte/ Honte en ce mont avoec poverte" (str. 41, vv. 484–85); as such it should be seen as an honour: "Dix, ki m'a donné de ses fleurs,/ Dist que tés hontes ert honneurs" (str. 39, vv. 466–67).

Noble though these sentiments are, the reality is harsh in this literal and permanent sentence of death to the world. Bodel says they may no longer converse with the healthy for fear of infecting them: "Car des or criem que ne lor nuise" (str. 1, v. 12); the hurt of this exile contained in the fact that as a leper he can no longer sit at table with the healthy: "Car ne puis mais nape tenir/ Entre sains puis que je mesale" (str. 8, vv. 95–96); and must take his leave of the most courteous of ladies, "la dame de Tenremonde" (str. 29, v. 465). Of greatest pathos in the case of Baude Fastoul is that he cannot protect the interests of his son and daughter (str. 54, vv. 637–48). Lepers must remove themselves from public paths when others use them: "Mes mals que je trai a tesmoing/ Fait que vous wide les sentiers" (Baude Fastoul, str. 5, vv. 56–57). Finally, in the last exercise of his most social of professions, Baude Fastoul laments that he is no longer a minstrel: "Car je sui uns hors menestreus" (str. 53, v. 636) – a compelling symbol of his death to the world.

There is an allegorical process at work here, side by side with realism. Heart (Cuers), Pity (Pitiez), Trouble (Anuis), are Jehan Bodel's messengers to the people from whom he takes his leave; and these personifications are taken up by Baude Fastoul. Also, Jehan Bodel conceives of his leprosy as a military process (vv. 74–78; 187; 188; 332; 357). He is entering an army (of the outcast lepers) from which none return: "Car d'aler en un ost m'atorne/ Dont nus en santé ne retorne" (str. 16, vv. 184–85); and he is caught in a net (perhaps that of the hunter rather than the soldier):

> Car en moi s'aüne et amasse
> Tous li anuis qui joie estaint–
> D'un mal de quoi nus ne respasse
> Puis que il a plain coup l'ataint (str. 22, vv. 260–64).

Pagans now have a firm truce from him: "Paien ont de moi ferme triue" (str. 29, v. 344), for his battle is now against his leprosy: "A parfournir ceste bataille/ Dont chascuns doit avoir pité" (str. 41, vv. 485–86).

In addition Jehan Bodel makes use of peasant life and games of chance, which last is also popular with Baude Fastoul, who additonally incorporates elements such as the English, fowling, politics, and the Last Judgement, as Pierre Ruelle points out on p. 78 of his edition. For Baude Fastoul, death is a tournament (v. 30); while leprosy is both a bouquet of flowers God gave him: "Dix, ki m'a donné de ses fleurs" (str. 39, v. 466) and a divine loan: "Ançois que Dix m'eüst presté/ Une enferté ki me fait mal" (str. 58, vv. 689–90). Thus we see that war is not the only image used to represent leprosy and the fate of the leper, but I emphasize it here because its timeless and universal nature has led it to survive, where the more parochial metaphors of these Arras poets have to a greater extent lost their meaning for us.

While the pictures painted by our poets are tragic, there is here no castigation of the sick for having the disease. However, one important concept we must consider here is the notion of shame. We have seen both men accept their affliction as coming from God to give them the chance to save their souls. However, neither poet seems imbued with any extraordinary sense of personal culpability over the evil done with which this expiation can be closely linked. Such statements smack more of a stereotypical stance of humility and acquiescence to God's will.

Yet both stricken poets express this idea of shame. Jehan Bodel tells us that his falling sick with leprosy is a shameful and ugly event: "Bauduïn Fastoul, or m'emplaide/ Une ochoisons honteuse et laide/ Qui m'a fait changier mon estage" (str. 28, vv. 325–27). He iterates the shameful element in his situation when he says that shame diminishes him: "Avec ce m'amenist et use/ Honte, que je criem et doute," (str. 13, vv. 152–53). He adds that shame and troubles assail him: "Que hontes et anuis m'enyvre,/ Qui nuit et jour assaut me livre" (str. 24, vv. 280–81); and that shame is always at his side: "Et honte, qui me reconvoie" (str. 31, v. 366). However, except in the first citation there is no connotation that he considers leprosy itself as shameful.

We hear similar statements from Baude Fastoul. Since he has become sick his body will be a source of shame to him until his soul surrenders: "En honte veut mon cors despendre/ Tant que l'ame li puisse rendre" (str. 9, vv. 103–04). This shame is his ration: "Hontes sera si li biaus don" (str. 13, v. 155); he lowers his head in shame: "Salue de par Beterel/ Cui hontes fait le front baissier" (str. 28, vv. 329–330); shame is always his companion: "Car maus et hontes me convoie" (str. 33, v. 389); and it rises to his very forehead: "Hontes, ki m'est montee u front" (str. 38, v. 454).

We learn that not only does leprosy destroy Baude, but shame is a considerable burden of itself: "Hontes, ki mon cors dessenist,/ Ki tout m'abat et asouplist" (str. 44, vv. 517–18). If we recall what Susan Sontag has said about the guilt attached in modern times to the sick: "Because of countless metaphoric flourishes that have made cancer synonymous with evil, having cancer has been experienced by many as shameful,"[60] this gives us pause for thought. She had suggested that the sick are made responsible for their sickness, but we see no hint of this in these medieval leprous poets beyond the general religious concept of atonement. We must look further to find the exact cause and nature of the leper-poets' shame.

Sontag also made the point that the most metaphorized diseases are in fact the most dehumanizing: "The most terrifying illnesses are those perceived not just as lethal but as dehumanizing, . . . the rabies phobia of nineteenth century France was caused by the fact that it turned people into maddened animals."[61] Here she may be closer to the mark. She had pointed to the leper's lion face with its progressive decomposition;[62] and indeed Baude Fastoul's comment that shame had risen to his forehead may be a reference to leprosy having attacked his face, because he openly says elsewhere that this is the case: "Puis que ma face est entamee/ Du mal ki n'est plaisans ne biaus" (str. 20, vv. 229–230). In one telling statement he admits that all disdain him because disease maims his body: "Enfertés ki mon cors mehaigne,/ Pour coi tous li mons me desdaigne" (str. 26, vv. 301–302). This comment strongly suggests that, in addition to the fear of contagion that obviously existed, the ugliness of the lesions of leprosy disgust the onlookers and hence cause shame to the sufferer.

Petrarch has also written on leprosy, presenting it as a shameful disease: "But what if unbearable pain is aggravated by a loathsome disease, nauseating and shameful? What if leprosy afflicts that pathetic, putrescent body?" (De remediis utriusque fortunae, Dialogue 114, "Severe Pain," 1.110).[63] Petrarch does not clarify here whether the shame goes beyond embarrassment at presenting an offensive spectacle to other people's gaze. However, it is perhaps implicit that this is the nature of the leper's shame when Petrarch adds that the disease is sent by God (1.110), and asks what good it does to add to the sickness of the body sickness of the mind (1.130); and when he further posits that defects of the mind are worse (1.150).

Some peoples have a low tolerance for handicapped and diseased fellow-humans, whom they consequently shun, and this may have been the case in Western medieval society. That this may be so is hinted at by both Jehan Bodel and Baude Fastoul's admission that they must henceforth cover their heads. Baude Fastoul says: "Raisons ne veut que je m'apere/ Plus au siecle cief descouvert" (str. 52, vv. 617–18); Jehan Bodel takes his leave ashamed and shrouded "Pren congié [of Thiebaut de la Pierre] honteus et couvers" (str. 5, v. 62). But he takes his leave, ashamed and head covered as one whom Fortune casts down: "Com cil qui fortune desmonte" (str. 5, v. 63). An antipathy towards physical ugliness does not of itself denote an allegorization of the disease. However, since Fortune was envisaged as all-powerful in the Middle Ages, an allegorization of God's divine Providence, Bodel's last statement cannot suggest guilt, but perhaps a feeling of embarrassment all may feel when struck by misfortune.

In the Middle Ages, leprosy was seen as a personal disease even though exiled lepers congregated together for support. Pestilence, however, was considered as a problem for society, the different attitudes deriving perhaps from the speed of contagion or infection of the latter. During this period, disease such as leprosy simply is, without connotations of cause and effect, as we have seen in the case of the lepers in Bèroul's *Tristan* and the leper poets. Plague, however, was often seen as a visitation by God. Susan Sontag reminds us of how in the ancient world disease was an instrument of divine wrath, with plague occurring in *Oedipus* because of a royal sinner.[64] Metaphorically, plague was the worst collective calamity, as Sontag reminds us when she recalls that "Procopius in his masterpiece of calumny, *The Secret History*, called the Emperor Justinian worse than the plague."[65] It is therefore not surprising to find that Christianity considered the concept of disease as punishment to be appropriate on occasion.

The Black Death appeared in Sicily before the end of 1347, spreading throughout Europe to cause unprecedented deaths. Boccaccio, describing its appearance in Florence in 1348, appropriates the concept of punishment when he writes that he does not know whether to ascribe the epidemic to the "heavenly bodies or our own iniquities which the just wrath of God sought to correct."[66] In all events the evident purpose of his description is to reveal how the Plague caused the Florentines to behave badly. Petrarch, who also laments the extent of the calamity, likewise wonders if this is the revenge of an angry God for their sins rather than a malign influence of the heavens or the effect of natural causes:

> Hoc gemuit Italie populus tot mortibus impar,
> hoc exhausta viris defectaque Gallia plorat,

> hoc alie quocunque iacent sub sidere gentes,
> Sive est ira dei, quod crimina nostra mereri
> certe ego crediderim, seu sola injuria celi,
> natura variante vices. Hic pestifer annus
> humano generi incubuit flendumque minature
> excidium mortique favet densissimus aer
> (*Epistole metriche* 7, "Ad seipsum," vv. 15–22).[67]

Philippe de Mézières gives precise descriptions of the maladies cured by the saintly offices of the dead Peter Thomas, as we have noted above, but in contrast he does not describe in detail the plague that swirled through Europe during his lifetime. His one-time master Lucchino Visconti, whom he served in 1345, died of the plague in 1349,[68] yet he makes no mention of this fact. In narrating the efficacity of Peter Thomas's prayers during the epidemic of plague on Cyprus, Philippe describes not a single symptom of this formidable disease.[69] Elsewhere he is equally laconic, writing summarily in his *Lettre sur la Présentation de la Vierge* that the ills of pestilence, sedition, deaths, wars, treachery and heresies had befallen the West.[70]

However, he does allegorize plague: in showing Peter Thomas preaching penitence as the remedy, Philippe makes of plague a scourge of God.[71] He presents this same image, when he narrates the sacrifice of Marcus Curtius, the Roman knight who leapt into the abyss astride his horse to save the Romans from the pestilence,[72] thereby indicating, as did his source St. Augustine in *De civitate Dei* (5.18), that since expiation would stop the epidemic, sin had caused it in the first place.[73]

Perhaps the failure to expound on the plague was due to the heavy mortality it caused, and particularly to the speed with which it killed. Philippe noted the propagation of the Roman plague by the air, before which there was no protection.[74] This is a trait to which to which he returns in his last work: *Epistre lamentable et consolatoire*, where he allegorizes the plague to explain that vices caused the defeat of the Christians at Nicopolis in the same way that this disease kills. Writing here in considerable confusion of the variant forms of bubonic plague, for it was surely the pneumonic form that killed so swiftly,[75] Philippe notes that as soon as the air is infected, the lump in the groin or armpit appears.[76]

While such confusion of symptoms denoted an unawareness that different types of plague existed, lack of precise medical knowledge did not impede medieval authors from establishing complex systems of medical allegory. Philippe de Mézières provides an illustration of this point: he presents the vices that assail the medieval woman's marriage vows and the virtues that protect them as a medical compendium of diseases and their cures.[77] Although he displays no extraordinary medical knowledge beyond that readily available to non-practictioners in his day, he presents a system. Taking as his point of departure the long-established theory of four humors and four conditions, where melancholy (or yellow bile) is cold and dry, choler (or black bile) is hot and dry, phlegm is cold and moist, and blood is hot and moist,[78] he presents a theory of health where the humors are present in the correct proportions, with disease occurring when there is a dominance of the one or a lack of the other. Seven principal maladies are caused by imbalances in the humors: dropsy is caused by corrupt blood and excess phlegm; paralysis and apoplexy come from black choler and corrupt phlegm; cholic passion is due to excess choler and corrupt vitelline (yellow bile); a cough comes from red and black choler, phlegm, and corrupt blood; frenzy is due to excess blood and burning choler, while heart failure and fainting arise from an excess of the four humors creating toxins. He identifies seven fevers: passing, slow, erratic, tertiary, quaternery, quotidian, and unremitting. For these maladies he prescribes seven medicines and electuraries: "dyaprunis, dyapenidon, dyalaxatif, dyaambra, dyaanthos, dyacention, and dyamargariton."

Relating each of these maladies and their treatment to conditions of the soul, we learn that the seven spiritual maladies spring from four conditions: disordered appetite, unbridled anger, uncontrolled self-will, and deep-seated jealousy, Thus dropsy represents the sin of pride, for the patient swells due to corrupt blood and over-abundant phlegm, i.e. she becomes swollen with pride and seeks dominion. Philippe teaches the etiology of these diseases: such maladies are engendered by lack of sufficient digestive heat coming from the liver to the stomach to cook food and provide good digestion. When heat is lost due to a surplus of cold humors, food is converted into corrupt humors and the stomach hecomes cold and maladies arise. The heat that keeps marriage healthy is love.

The art of astrology is also included. Traditionally, each of the humors is associated with a planet: melancholy with Saturn, choler with Mars, phlegm with the Moon, and blood with Venus.[79] Philippe associates each of his maladies with one of the seven planets and specifically indicates that a particular medicine is to be administered under the ascendence of a particular planet, and this in conjunction with the appropriation of one of the seven metals of the earth (such as gold, tin, lead, mercury) and one of the seven precious stones (such as the emerald, diamond, ruby and orient pearl). In his concordance, Philippe equates the carbuncle with Christ and the diamond with the Virgin, while the white, round, shining orient pearl is the wafer of the Eucharist. In addition, his moral prescription includes the admonition to the patient to take appropriately from the seven gifts of the Spirit, the seven sacraments, the seven works of mercy, and the moral and theological virtues.

Philippe carries his symbolism further, allegorizing the compass (used at that time by sailors in the Adriatic but not in the North Sea, as he tells us) to represent the human body.[80] He describes the navigator's box as having a point of metal in the middle on which is seated a flat, round card, able to turn like a wheel, although independently of the box. Lines painted on the card represent the winds, the four main ones being the East, West, North and South winds. Attached to the card is an iron needle with the point fixed on the line representing the North Wind. To use this box one rubs the point of the needle with the lodestone and the needle points to the polar star by which sailors chart their path.

The box, Philippe explains, is the human body. The wheel and the four winds are the parts of the body: in an apt parallel with the four conditions deriving from the four humors, the East Wind, which is hot and wet, is the head; the West, which is hot and wet, is the legs and feet; the South Wind, which is hot and dry, is the right arm, shoulder, and hand; while the North Wind, which is cold and wet, is the left arm, shoulder, and hand. The other winds are the internal organs: the brain, stomach, liver, and lungs, and the four complexions which go from one to the other. The needle is the heart and the point of metal on which the needle sits is human life, which, Philippe poignantly points out, is only a little point of rotten and failing metal. The lodestone is the Virgin, representing her humanity in this world, while the polar star is the Queen of Heaven. Thus does Philippe incorporate advances in technology to expand St. Bernard's great metaphor of the Virgin as the Star of the Sea.[81]

It is clear from the foregoing that one principal difference between medieval and modern allegorical treatments of disease is that where modern medical comparisons present limited metaphors, such as viruses attacking computers and crime as a cancer in a city, medieval allegory was detailed, comprehensive and, above all, systematized. The other main difference derives from the figure of the physician. While the modern physician enjoys an inflated status, he or she is a pale shadow beside the medieval figure, for medieval medical allegory posited The Physician, Christ, at its center. While a disease such as leprosy was a death sentence, the basic image was not one born of empty despair and intended to create horror, for disease was seen as coming from God and was invested with meaning. Because there was Christ, who performed miracles of healing and raised the dead to life, such allegory offered a comprehension of God's promise of eternal life. Thus the allegorical process is one of the ways whereby a mystic can verbalize mystical experiences, and we have seen the Middle Ages use it as such.

University of Long Island
New York

NOTES

[1] *The Physician in Literature*, ed. and intro. Norman Cousins (Philadelphia: Saunders Press, 1982, p. xv).

[2] *The Works of Geoffrey Chaucer*, 2d ed. Ed. F. N. Robinson (Oxford: Oxford UP, 1957).

[3] *Le Songe du vieil pelerin*, ed. George W. Coopland, 2 vols. (Cambridge: Cambridge UP, 1969), 2:321–22.

[4] Philippe de Mézières, *Le Songe* 2:272.

[5] Philippe de Mézières, *Le Songe* 2:273.

[6] *Oeuvres complètes d'Eustache Deschamps*, ed. Marquis de Queux de Saint-Hilaire and Gaston Raynaud, SATF (Paris: Firmin Didot, 1880), 2:226–235.

[7] Isidore of Seville, *Eytmologies* Bk. 4.3.1–2.

[8] Frederick F. Cartwright and Michael D. Biddis, *Disease and History* (New York: Thomas Y. Crowell, 1972), p. 20.

[9] Cartwright and Biddis, *Disease*, pp. 20–23.

[10] Joachim Gruber, *Kommentar zu Boethius: De consolatione philosophiae* (Berlin: Walter de Gruyter, 1978), pp. 32–38.

[11] An excellent treatment of this movement, and the source of my information here, is contained in Conrad H. Rawski's translation and commentary on Petrarch's *De remediis utriusque fortunae*, Commentary to Bk 1, vol. 2, p. lxi-lxii, n. 19 to "Dramatis personae: The Argument of the Book." (*Petrarch's Remedy for Fortune Fair and Foul*, tr. and commentary by Conrad H. Rawski, Bloomington, Ind.: Indiana UP, 1991, 5 vols.).

[12] Hugh de Saint-Victor, *De Medicina animae* P.L., vol. 176, cols. 1183–1202 (1854, Turnhout, Belgium: Brepols, 1983).

[13] In *Song of Songs* I, tr. by Kilian Walsh OCSO, Cistercian Fathers Series **4** (Kalamazoo, Mich.: Cisterican Publications, Inc. 1981), p. 1.

[14] *The Golden Legend of Jacobus de Voragine*, tr. and adapted by Granger Ryan and Helmut Ripperger, 2 parts (London: Longmans, Green, 1941), 2:577–78.
[15] Philippe de Mézières, *The Life of Saint Peter Thomas*, ed. Joachim Smet, O. Carm (Rome: Institutum Carmelitanum, 1954), p. 171.
[16] This was the name given to tubercular adenitis, i.e. inflammation of the lymph nodes due to the bacillus of turberculosis, which caused suppurating, putrid sores, particularly of the face; and affected any of the lymph nodes, including those of the groin, thereby mimicking bubonic plague.
[17] Marc Bloch, *The Royal Touch: Sacred Monarchy and Scrofula in England and France*, tr. J. E. Anderson (London: Routledge and Kegan Paul, 1973), pp. 11–12. (Tr. from *Les Rois thaumaturges* [Paris: Armand Colin, 1961].)
[18] Paris, Bibliothèque Nationale, MS. Lat. 2900, f. 14, published in P. L., vol. 156, col. 616. See Bloch, *The Royal Touch*, pp. 12–14 and nn. 8 and 11 to chap. 1.
[19] Paris, Bibliothèque Nationale, new acq. MS. Lat. 785, f. 59, published in P. L., vol. 207, col. 440 D. See Bloch, *The Royal Touch*, pp. 21–22 and n. 25 to Chap. 1.
[20] William of Malmesbury, *Historia Regum*, ii, I, 222, ed. Stubbs (Rolls Series, i.) p. 273. See Bloch, *The Royal Touch*, pp. 22–25 and n. 33 to chap. 1.
[21] *Le Songe*, 22:250.
[22] *Le Songe*, 2:132.
[23] *Le Songe*, 2:132–33.
[24] New York: Farrar, Straus and Giroux, 1977, and New York: Farar, Straus and Giroux, 1988, respectively.
[25] *AIDS*, p. 26.
[26] *AIDS*, p. 27.
[27] "Reluctantly, Black Churches Confront AIDS," p. 1, cont. on p. B8, col 1.
[28] National News Section, p. 16.
[29] *AIDS*, p. 23.
[30] *Illness*, pp. 12–13.
[31] *Illness*, p. 14.
[32] *Illness*, p. 117.
[33] *Illness*, pp. 17–18.
[34] *Illness*, p. 20.
[35] *Illness*, p. 18.
[36] *Illness*, pp. 14–15.
[37] *Illness*, pp. 16–17.
[38] *Illness*, p. 57.
[39] Sontag, *AIDS*, p. 24.
[40] *AIDS*, p. 38.
[41] *AIDS*, p. 94.
[42] *La Vie de saint Louis. Le témoignage de Jehan, seigneur de Joinville*. Texte de XIVe siècle, ed. Noel L. Corbett (Sherbrooke, Quebec: Naaman, 1977). Paragraphs 27 and 28 are devoted to this discussion of mortal sin and leprosy.
[43] Saul Nathaniel Brody, *The Disease of the Soul: Leprosy in Medieval Literature* (Ithaca, N.Y.: Cornell UP, 1974), pp. 64–67.
[44] *Illness*, p. 58.
[45] Guy de Chauliac, the papal physician, wrote in 1363 that leprosy is a very injurious disease, that it is hereditary and contagious, and almost impossible to eradicate (*La Grande Chirurgie de Guy de Chauliac*. ed. E. Nicaise [Paris: F. Alcan, 1890] p. 406).

[46] Brody, *The Disease of the Soul*, p. 60.
[47] Thomas, *Les Fragments du Roman de Tristan*, ed. Bartina H. Wind, Textes littéraires français (Geneva: Droz, 1960). There is a second instance of Tristan assuming the disguise of a leper: he awaits Iseut at the ford so that, having carried her on his back, she can truthfully swear that no man other than her lord and this leper has been between her legs (Béroul, *Le Roman De Tristan*, ed. Ernest Muret, 4th ed. rev. L. M. Defourques [Paris: Honoré Champion, 1967], vv. 3563–984), as Merritt R. Blakeslee points out (*Love's Masks: Identity, Intertextuality, and Meaning in the Old French Tristan Poems* [Cambridge: D. S. Brewer, 1989], p. 67).
[48] Béroul, *Le Roman de Tristan*.
[49] Frederick F. Cartwright and Michael D. Biddis, *Disease*, p. 60.
[50] Brody, *The Disease of the Soul*, pp. 107–146.
[51] Peter Richards, *The Medieval Leper and His Northern Heirs* (Cambridge: D. S. Brewer, 1977), p. 6.
[52] Peter Richards, *The Medieval Leper and His Northern Heirs* (Cambridge: D. S. Brewer, 1977), p. 5.
[53] Brody, *The Disease of the Soul*, pp. 51–52, and n. 52.
[54] R. R. Burg, *Sodomy and the Perception of Evil: English Sea Rovers in the Seventeenth-Century Caribbean* (New York: New York UP, 1983).
[55] *AIDS*, pp. 59–60.
[56] Pierre Ruelle, *Les Congés d'Arras (Jean Bodel, Baude Fastoul, Adam de la Halle)*, Travaux de la Faculté de Philosophie et Lettres **27** (Bruxelles: Presses Universitaires de Bruxelles and Presses Universitaires de France, 1965).
[57] Ruelle, *Les Congés*, p. 60.
[58] Ruelle, *Les Congés*, p. 68.
[59] *AIDS*, p. 38.
[60] *AIDS*, p. 24.
[61] *AIDS*, pp. 38–39.
[62] *AIDS*, p. 41.
[63] Petrarch's *Remedies*, vol. 3.
[64] *Illness*, pp. 39–40.
[65] *AIDS*, p. 44.
[66] *The Decameron*, tr. Frances Winwar (New York: Modern Library, 1930), p. xxiii.
[67] Francesco Petrarca, *Rime. Trionfi. E Poesie latine*, ed. F. Neri, G. Martellotti, E. Bianchi. N. Sapegno (Milan/Naples: Riccardo Ricciardi, 1951), p. 750–759.
[68] Aziz S. Atiya, *The Crusade in The Later Middle Ages* (London: Methuen, 1938, rpt. New York: Kraus, 1965), p. 137.
[69] Smet, *The Life of Saint Peter Thomas*, pp. 7–100.
[70] William E. Coleman, *Philippe de Mézières' Campaign for the Feast of Mary's Presentation*, Toronto Medieval Latin Texts (Toronto: Pontificial Institute of Mediaeval Studies, 1981), p. 41.
[71] Smet, *The Life of Saint Peter Thomas*, p. 100.
[72] *Letter to King Richard II*, ed. and tr. George W. Coopland (Liverpool: Liverpool UP, 1975), pp. 95–96.
[73] This image must have struck the fourteenth century imagination, for Petrarch also refers to this incident in his "De Remediis utriusque fortunae," Bk. 1 dialogue 90, "Kingship and Empire," 1.209, Petrarch's *Remedies*, 1:263.

[74] *Letter to Richard II*, p. 95.

[75] The Black Death of 1346–61 is defined as a pandemic of the pneumonic type of bubonic playue by Frederick F. Cartwright and Michael D. Biddis in *Disease*, p. 31.

[76] "L'Epistre lamentable et consolatoire sur le fait de la desconfiture lacrimable du noble et vaillant roy de Honguerie par les turcs devant la ville de Nicopoli," partially published in vol. 16 (pp. 414–523) of *Oeuvres de Froissart: Chroniques*, ed. Henri Marie Bruno Joseph Léon Kervyn de Lettenhove (Brussels: V. Devaux, 1872), p. 518.

[77] "Le Livre de la vertu du mariage du sacrement de mariage," Bk. 3. This text is contained in Paris, Bibliothèque Nationale, MS. fr. 1175, an edition of which I am completing.

[78] Maurice Hussey, *Chaucer's World: A Pictorial Companion* (Cambridge U.K.: Cambridge UP, 1927, rpt. 1967), p. 105.

[79] Hussey, *Chaucer's World*, p. 105.

[80] *Le Livre*, Bk. 3, chap. 25.

[81] Bernard of Clairvaux and Amadeus of Lausanne, "Magnificat. Homilies in Praise of the Blessed Virgin," tr. Marie-Bernard Saïd Perigo, Cistercian Publications **18** (Kalamazoo, Mich.: Cistercian Publications, 1979), *Magnificat* 1.17.1.

HEATHER ROSS

MICHAEL DRAYTON'S 'IDEAS' AND THE 'WHERE' AND THE 'WHENCE' OF ALLEGORY

Although still in print and on the shelves of libraries, most of the poetry of Michael Drayton is no longer either read or taught. If the canon of English Literature has preserved anything of him, it has saved not him but one poem (and an extremely powerful one): a mere sonnet, a mere fourteen lines, "Since there's no helpe, come let us kisse and part."[1] For our purposes here, it is significant that Drayton lived from 1563–1631 and that, hence, his life overlapped with that of the great allegorist, Spenser, and extended well beyond that time: 1631 was also the year that John Donne died. Allegory, by then, was no longer in fashion.

Four of Drayton's collections of poems contain the symptomatic word "Idea" in their titles: *Idea. The Shephred's Garland Fashioned in Nine Eglogs* (1593), *Idea's Mirror. Amours in Quartorzains* – a collection of sonnets (1954), *Endymion and Phoebe. Idea's Latmus* – an epyllion, or Ovidian narrative poem (1595), and *Idea. In Sixty-three Sonnets* (1619). Interesting for what it may say about the relationships between vision and revision, the sonnet sequences which began in 1594 underwent drastic rewriting and eleven new editions (1599, 1600, 1602, 1603, for instance) until 1619. During this time (a span of twenty-five years), Drayton clearly became his own best reader, reproducing meaning and hence, rewriting with each new reading. My approach to Drayton here is two-fold. First I want to argue that Michael Drayton was not only a practitioner of allegory but also a complex (and, perhaps, at times nostalgic) thinker about the nature of allegory. My second suggestion, perhaps more than a suggestion, is that Drayton is a pre-Husserlian phenomenologist and a prophet of phenomenlogy's consequence for literary theory: hermeneutical phenomenology and reception theory.

It is Angus Fletcher[2] who most approximates a plausible reason for the failure of Drayton's poetry to survive: "it demanded an exegetical labor that was already out of fashion," and he quotes Drayton himself in an address to the reader of one poem: "If, as I say, thou hadst rather (because it asks thy labour) remain where thou wert, than strain thyself to walk forth with the Muses, the fault proceeds from thy idleness, not

M. Kronegger and A-T. Tymieniecka (eds.), Analecta Husserliana XLII, pp. 83–91.

from any want of my industry." Perhaps it is the impatience and crankiness of this 1613 comment which led to the biting sarcasm of Oliver Goldsmith in a 1760 letter to *The Citizen of the World*:

As we walked along to a particular part of the temple, There, says the gentleman, pointing with his finger, that is the Poet's Corner there you see the monuments of Shakespeare, and Milton, and Prior, and Drayton. Drayton! I replied, I never heard of him before; but I have been told of one Pope; is he there? It is time enough, replied my guide, these hundred years; he is not long dead; people have not done hating him yet.[3]

But before his 1613 tantrum with his reader/readers, Drayton has some remarkably modern exhortations to the reader, suggesting that the object of the poem (the text) does not exist without subject of the reader, producing the test. "Letters and lines," he begins in "Amour 21," "we see are soon defaced": "paper and ink can point but naked words." Sighs, he concludes in the octet, "and even signs a silly hope affords." The *volte* (turn) of the sonnet in the sestet is characterized by an apostrophe (literally, a "turning away") to a personified figure, "Shadow," in whom, Drayton trusts, the reader (subject) will find his own shadow (as object): "That everything whence shadow doth proceed/ May in his shadow my love's story read." Consistent with his continuing concern with the reader, the final 1619 version of Drayton's *Idea* begins with a sonnet-caveat to his reader: "Into these loves who but for passion looks,/ At this first sight here let him lay them by,/ And seek elsewhere, in turning other books." A libertine, he calls himself, "to variety inclin'd": "My verse is the true image of my mind,/ Ever in motion, still desiring change."

This inauguratory sonnet which begins *Idea* is, of course, totally ironic: what the reader will find, indeed, is the passion and the "ah-me's" from which the poet dissociates himself. What interests me most here about Drayton's use of irony in *Idea* is its theoretical similarity with allegory. The traditional definition of verbal irony is that the speaker's implicit meaning differs sharply from the meaning that is ostensibly expressed.[4] Similarly, a traditional starting point for a definition of allegory is "speaking otherwise than one seems to speak" or "saying" one thing and "meaning" another.[5] As M. H. Abrams underscores, irony, such as that found in Drayton's first sonnet, "carries an implicit compliment to the intelligence of its readers" by inviting them to become one with "the author and the knowing minority who are not taken in by the ostensible meaning."[6]

That the world of Drayton's sonnets is populated with personified figures ("the dreary abstracts of my endless cares") is uncontestable: Love, Time, Night, Death. Passion, Faith, Innocence inhabit and, indeed, control the domain of his Ideas. Allegory, a term associated almost inevitable with narrative rather than the lyric, or sonnet, may seem a less appropriate term than Personification for Drayton's poetic employment of such abstractions. Apostrophe ("Stay, stay, sweet Time") is also unmistakably a rhetorical device which is extremely active in his sonnets. Certainly, allegory begins with personification; and what make Drayton's allegory difficult is that he tends to shift comfortable back and forth between a narrative mode of history, time, situation, impossibly fact, to a mode which represents the abstractions of Passion and Faith, for instance, as if what is sometimes called the "vertical" world of allegory could co-exist and cooperate in the horizontal world of the narrative which is not fiction but history.

In the past three decades, theories of the nature of allegory have ranged from the general (and, hence, sublime) to the very particular (and, hence, grammatical and rhetorical). In 1958, C. S. Lewis wrote that allegory in some sense, "belongs not to medieval man, but to man, or even to the mind, in general."[7] It is the very nature of thought and language to represent what is immaterial in picturable terms. Angus Fletcher, who first used the "says one thing and means another" definition of allegory in 1964, speaks both of the linguistic implications (Allegory destroys the normal expectation we have about language, that our words "mean what they say")[8] and the spiritual (the apocalyptic escape into infinite time and space: allegory is serving major social and spiritual needs).[9] similarly, Maureen Quilligan in 1979 speaks eloquently of the power of allegorical texts: they are always fundamentally about language and the ways in which language itself can reveal to man his highest spiritual purpose in the cosmos . . . its existence assumes an attitude in which abstract nouns not only name universals that are real, but in which abstract names themselves are perceived to be as real and as powerful as the things named.[10] Even J. Hillis Miller, drawing a parallel between parable and allegory in 1981, speaks uncharacteristically: Word becomes flesh in allegory – Man, society, and nature, which seem so solid and self-subsistent, are made and unmade by spiritual forces embodied in symbols.[11] Less euphorically, Murray Krieger concentrates (again in 1981) on language and semiotics in allegory, regretting that in it we see the signifier looking helplessly across the chasm of time at an unreachable signified,[12] and he agrees with de Man: allegory designates primarily a distance in relation to its own origin.[13] Quite correctly, I think, and quite specifically, Samuel R. Levin clarifies what seems to have become the operative definition of allegory both lucidly and specifically: when it is claimed that allegory says one thing and means another, it is the predicates that "say" the one and the (personified) nouns that "mean" the other.[14]

To the converted, then, allegory has something to do with an affair between consciousness and things divine. To the unbelievers, it is a rhetorical trope, a wrenching of a word and words from their everyday meaning. Cross-references in Classical and Renaissance rhetoric define allegory as *inversio* and connect it with a variety of other tropes: *permutatio* (change, substitution) for a start. But *enallage* (a shift in person or gender, for instance) is also a form of *permutatio* and is, in turn, linked to some of the vices of language (*anthimeria* and *antiplosis*) as well as the legitimate and frequently used figure of repetition (a scheme rather than a trope) *anadiplosis* (or duplication, used frequently in the creation of rhetorical climax).[15] For one Renaissance rhetorician (Peacham), allegory is defined as *enigma*.[16] Hinted at occasionally is its connection with *ekphrasis* (description) as well as with irony and both *metonymy* and *synecdoche*.[17]

My own stance with respect to the linguistic and rhetorical questions opened by allegory is two-fold. First, as it works in Drayton anyway, allegory may very well be defined as the teleology of synecdoche (the part standing for the whole): my love stands for our or all love, and finally, the allegorical figure of Love. Secondly, Drayton's use of allegory is not unconnected with his use of irony (both are linguistic permutations); and if it is the predicates which say one thing and the nouns that mean another in allegory, it is the nouns which says one thing and the predicates which mean another in irony. It is with more than a little trepidation that I approach the subject of the sign (of the signifier and the signified) since now, 1992, the signifier is almost universally understood to be nothing but a sound or a mark on the page and the signified is only the hypothetical matching concept in the mind. To Drayton, perhaps, signifiers were what existed in the world of history and in the world of time whereas, perhaps, the signifieds dwelt in a realm beyond time and history (where Endymion sleeps and the poets dream). For Drayton and, I suspect, for much of his age, signifier and signified were manufactured together in the subject-object productions of mind and conciousness which both was and is poetry. The interrelationship between allegory and history, between history both and (and as) allegory, between signifier and signified is, in a Thomistic sense, relational: one term cannot exist without the other. As there can be no Master without a Servant, Prince without a Subject, or Son without a Father, so too there can be no Signifier without a Signified.[18]

Now to the text which has survived the name of Drayton, "Since there's no help, come, let us kiss and part." Most of us have "been there": love (and passion) still exist but the relationship, presumably impossible, cannot allow for the continuance of love in the world of time – "not in this lifetime," as the saying goes. The situation is ineluctable. The octet, the first part of the poem, is highly rhetorical (an attempt to persuade both and the other that it's over): "And I am glad, yea glad with all my heart." By the second quatrain of the poem the speaker is impatiently imperative: "Shake hands forever, cancel all our vows." The *anadiplosis* of line three ("I am glad, yea glad with all my heart") is an indicator of the building climax which culminates at the end of the octet with the death (the ultimate climax) of love: "And when we meet at any time again/ Be it not seen in either of our brows/ That we one jot of former love retain." Since my subject here is allegory and, to some extent, irony, where on both cases one thing is said and another thing meant, one cannot help but comment about the octet that we drive home the power of the poem is that one thing is said (perhaps must be said) and that two things are meant. On the one hand the speaker means what he says when he says "I've had it" (Drayton's "Nay, I have done, you get no more of me") and, on the other hand, what is also meant is the optative opposite of what seems most to be meant: "It should not be so; I wish it were not so; I do not want to but am forced, by circumstance, to say so." Neither allegorical nor ironic at this point, the poem still cannot be said to say what it means. In fact, the protestation especially, "Nay, I have done," opens the entire poem to a deconstructive reading. And in many ways, one might well argue, what the poem is finally about is the impossibility of saying what one means.

A deconstructive reading of the poem would, of course, "read the poem against the grain," against its stated rather than intended purpose. The more the speaker protests, "Nay," the more he suggests that what he really means is the opposite of what he says. The poem, then, becomes a rhetorical exercise: an attempt to persuade both himself and others of what he most distinctly, originally, did not mean. From the very start he did not say what he "meant" but spoke with the manipulative power of a rhetorician whose language is by no means sincere but spoken with affective and final causes as its determining dynamic.

With the turn into the sestet, the speaker modulates into the discourse of allegory, presenting a poignant and communal description of Love, Passion, Faith, and Innocence together dying: "Now at the last gasp of Love's latest breath,/ When, his pulse failing, Passion speechless lies,/ When Faith is keeling by his bed of death/ And Innocence is closing up his eyes." The "now when" idiom presses home the immediacy and, indeed, urgency, of the situation (the death of Love), and the parataxis of the style in this quatrain contrasts itself to the hypotactic (cause-and-effect) style of the octet ("Since there's no help"). The contrast between hypotaxis, the hypotaxis of history and the parataxis of allegory suggests, moreover, the telelogical difference between the effective and final cause of things: allegory must be seen from the perspective not of effective but of final cause.[19] What final couplet does ("Now if thou would'st, when all have given him over,/ From death to life thou might'st him yet recover") is to invite the reader-addressee to put into effect the final cause, to resuscitate the abstractions of consciousness and transcendental signified and put them back into the world of history. Such is the project which Drayton asks of his reader: to give back to history the power of allegory.

My suggestion for understanding the poem, then, is to remove it from the deeply personal level in which it seems to be written and to which it is all too easy to respond and to think of it is as a poem written to all readers, suggesting that the fate of allegory as a living mode of expression is in the hands (or mind) of the reader. It is in the power of the reader to revive the dying figure of Allegory, the world of abstractions, and the fading figures of a land once inhabited by the gods.

That the poem is in part about power is self-evident: the speaker begins by assuming rhetorical power and ends by empowering his reader. Since much of the power struggle is ostensibly the struggle between male and female for power, it is necessary that one notice the inversed gender of Drayton's abstraction of Love, Passion, Faith, and Innocence: most atypically, they are all male, or are all one male. All the abstractions of Latin literature and after are engendered feminine: Fortune is *Fortuna*, etc. Maurren Quilligan, for one, refers, but not at length, to "the problems of female authority" which are "inherent in the linguistic basis of allegory."[20] Here, in the gender change, is an example of *enallage*: Love and Faith (Spenser's *Fedelia*), for instance, are improperly gendered. This fact, I think, is not to be taken lightly. Is it not further evidence of the telelogical synecdoche of allegory? The "he" who loves becomes "all-he" who loves, and this substitution is the ultimate permutation as well as inversion (of gender).

But such a striking inversion of the usual female gender of allegorical figures asks for no simple or single interpretation. Indeed, Love, and Faith, and Passion are all "he" in Drayton's poem, but they are also all "he's" who are lacking in male power and, indeed, in need of female power – the power to bring them back to life. What is especially curious is that in engendering his allegorical figures as male, Allegory loses its power, suggesting that allegory is not simply the product of the mind at all but is linked with the mind's relationship with the Other.

In his chapter on "Phenomenology, Hermeneutics, Reception Theory" in *Literary Theory*, Terry Eagleton objects throughout to the claims of Husserl, Heidegger, Gadamer, and Iser, among others: "Hermeneutics," he says, "sees history as a living dialogue between past, present, and future." But, he complains, the "dialogue of human history is as often as not a monologue between the powerful and powerless" – "it refuses to recognize that discourse is always caught up with power which may be by no means benign."[21] The powers to which Eagleton refers are, of course, ideological, the power of the oppressor over the oppressed. What Eagleton's own theoretical stance does not take into account is the fact of another power, perhaps greater than, or, at least, equal to the political power, and that is the power of eroticism, the power of desire, the desire for another, of another, to be in and with another, to lose and find oneself in another. Through eroticism power is given to, as well as taken from, another. The power which exists between the mind and the world around it, and which exists between allegory and history, is based on the model of erotic power. The story which it tells is finally a story about how it is through becoming another that we say one thing and mean the other.

Ideas Latmus, or *Endymion and Phoebe*, is an Ovidian erotic poem, and a poem about poetry and allegory. Time may be the stuff of consciousness, but space is more specifically the location of history. *Endymion and Phoebe* is filled from the very start with the lexemes of place and space: "whence," "whereon," "where," "under," "within," and "place" all appear at least once in the first twenty-two lines of the poem. What the poem begins by doing, then, is to give place and space to the

Ideas which are putatively of the mind: it is the place from which, according to the poem, language, idea, and poets "derive." Anachronistically, the poem answers de Man's charge that allegory designates primarily a distance in relation to its own origin. *Endymion and Phoebe* trace those origins of allegory through the power of the erotic: an Ovidian story told of the desire of the gods (or Ideas) for humans (for their bodies) and the sometimes at first reluctant desire of humans to live and with by the gods (and Ideas). What the poem deals with, then, is the attraction between Endymion and Phoebe, between the poet and Phoebus and Phoebe, as well as the poet's search for the abstractions which rule his destiny: Love, Passion, and Faith.

The relationship between abstraction concepts and the place-space inhabited by man and mind is, according to Ovid, metamorphic: human identity turns into object, and gods (or Ideas) turn into beasts who desire and seduce human identity.[22] Following the conflicting traditions of Plato and Ovid, then (a strange marriage), Michael Drayton understands allegory as an erotic affair between Idea and form, consciousness and history, with each one desiring, wanting, and even, needing the other to fulfill each other.

York University
Ontario

NOTES

1 All references to the poetry of Michael Drayton are to the Muses' Library edited by John Buxton (Harvard: Cambridge, Mass., 1953).

2 Angus Fletcher, *Allegory: The Theory of a Symbolic Mode* (Cornell: Ithaca, 1964), p. 235n.

3 Quoted by John Buxton in the Introduction to his edition of the poems of Drayton, p. xli.

4 See, for instance, M. H. Abrams, *A Glossary of Literary Terms*, 5th edition (Holt, Rinehart, Winston, 1988).

5 Angus Fletcher, p. 2.

6 M. H. Abrams, p. 92.

7 C. S. Lewis, *The Allegory of Love* (Oxford: New York, 1958), p. 44.

8 Fletcher, p. 2.

9 Fletcher, p. 23.

10 *The Language of Allegory: Defining the Genre* (Cornell: Ithaca, 1979), p. 156.

11 J. Hillis Miller, "The Two Allegories," in *Allegory, Myth, and Symbol*, ed. Morton W. Bloomfield, Harvard English Studies, No. 9 (Cambridge, Mass., 1981), p. 361,

p. 368. Such references to Word becoming Flesh I consider to be a major deviation from the principles of Deconstruction; hence, I say, "uncharacteristically."

[12] Krieger, "A Waking Dream," in Bloomfield, p. 4.

[13] Krieger, pp. 14–15.

[14] "Allegorical Language" in Bloomfield, p. 25.

[15] See the especially useful *A Handlist of Rhetorical Terms* by Richard A. Lanham, Second Edition (California, 1991).

[16] See Fletcher, P. 7.

[17] Again, see Fletcher, p. 85.

[18] *The Pocket Aquinas*, ed. Vernon J. Bourke (Washington Square Press: New York, 1960).

[19] My own language here is, once again, theological. The effective cause of anything is what comes prior to the thing caused: the effective cause of my own creation was the moment of my conception. The final cause of my creation is still to be discovered: the final cause is in the thought of the First Mover and involves not how I began but for what end will I become.

[20] Maureen Quilligan, *Milton's Spenser: The Politics of Reading* (Cornell: Ithaca, 1983), p. 172 and passim.

[21] *Literary Theory: An Introduction* (Minneapolis: University of Minnesota Press, 1983), p. 73.

[22] See especially Leonard Barkan, *The Gods Made Flesh* (Yale: New Haven, 1986), Chapter V, for an extremely interesting treatment of Ovidianism in the Renaissance. See, too, Clark Hulse, *Metamorphic Verse* (Princeton, 1981), particularly Chapter V.

PART II

Anna-Teresa Tymieniecka, Bill Smith, and Christine Raffini at the Institute's reception during the April Allegory Congress in Cambridge/Belmont 1992.

MARIE-FRANCE WAGNER

LES CONQUÊTES DE LA TOISON D'OR OF 1661 AND 1683: *The Breakdown of the Sun Allegory*

> Allegory is the most ancient device for expressing thought through the written word. Writing came into use long after the Egyptians left their mysterious figures as evidence of their religion and laws of state. Poets since that time, through the use of allegorical meaning, metamorphosis and fable, have handed down to us the finest precepts and most valuable lessons in philosophy [. . .]. And in the end, however little we may know of the poets themselves, we can understand their subject matter with no difficulty; for allegory, far from obscuring their ideas, enlightens and enriches them [. . .]
>
> *Mercure galant*, 1684[1]

Louis XIV was by no means the first to use the Sun as a symbol – from time immemorial the sun has been taken as a royal emblem. In Renaissance culture, the metaphor of the sun was used to characterize the role of the sovereign and his assembly. In 1537 Jacques Cappel wrote of King Francis I: "You are as the Sun among the planets,"[2] and he secularized the mediaeval *Topos* of the divine court of God, transforming it into a royal court on earth. Henry IV was the first "Sun King," and indeed in 1610 the president of the Parlement, de Harlay, said in his address to the newly-crowned young monarch that: "There is not a loss of that great sun, but rather an eclipse of it whereby its light, no sooner obscured in one place, appears in another."[3] These metaphors of phoenix and sun[4] served to illustrate the transition of power after Henry IV's death: the king's demise was shown in the clouds that dimmed the brightness of the sun, but that sun would shine once more, in the shape of Louis XIII.

It was in 1662 that Louis XIV officially adopted the sun, a codified image, as his central heraldic device. He chose as his allegorical mask the figure of Apollo: god of light, who triumphed over the monstrous

M. Kronegger and A-T. Tymieniecka (eds.), Analecta Husserliana XLII, pp. 95–112.

snake Python,[5] the god of potent virility, victor over chaos and the forces of the moon, choirmaster of the Muses, the god of healing and health, who defeated ignorance and barbarity. He explained his choice in his *Mémoires*:[6]

By his very uniqueness, by his dazzling brightness and the radiance he casts over the other stars which form a sort of royal court about him, by his just and equal manner of distributing this radiance among all the different regions of the world, by the good he brings to all places, bestowing life, joy and activity wherever he passes, by his endless movement across the heavens through which he somehow still remains serene, in the constant and unvarying course from which he never strays, the sun is undoubtedly the most vivid and beautiful image there can be of a great king.

By these considered words, "the monarch demonstrated his desire to see the sun image no longer simply applied to his person as some random symbol he had borrowed; he wished to establish it as the reflection of himself."[7] In 1666, Père Le Moyne, in his *Art de régner*,[8] presented the sun as "mirror and model of the king," and added that: "as well as being, as everyone knows, to the natural World what the Prince is to the Political World, the Sun is also the most fitting Symbol to represent the qualities and functions of a good Prince." This double function of imitation and substitution coexists in paradoxical fashion in every representation. In terms of the conventional cultural symbolic system, the sun is depicted almost as the emblem of the king.

In his *Toison d'or* of 1661, reprised in 1683, Corneille showed great vision in his staging of the customary sun figure. The Suns of 1661 and 1683 appear only once, with great effect, at the end of the play, when they emerge in a blaze of glory. The fable is introduced by two different prologues, each with something of the events of the day: in 1661, "on the marriage of His Majesty and the peace with Spain," and in 1683, on "the fruit of the union celebrated in 1661 now being honoured," a new prologue written by La Chapelle. These facts convey a snippet of contemporary history, but the celebration of a princely marriage and a royal birth, even as they helped perpetuate the monarchy, relegated Clio to the genre of eulogy.

The two *corpora* of the tragedy – the prologue (directly readable from the historical events) and the fable (the founding discourse, its meaning never completely fixed, for fable is the matrix, awaiting its potential representation by the artist) – undergo two complementary principles. On the one hand, we see the "mythologization" of history in the prologue, and on the other, the "historicization" of the fable, with the consequent production of a new text which encompasses the legendary intertext of prologue and polysemic myth. Thus the sun symbol is transformed into allegory – according to Furetière's definition, a "rhetorical figure which is a sustained metaphor, when a writer uses a certain discourse which is proper to one thing in order to convey something else." An allegorical reading is thus a conceptual reading: "the idea takes on body, becomes image, in order to touch our hearts and stir our senses, and, through pleasure, invites us to action."[9] Faithful to the Platonic spirit which indissolubly links image and idea, allegory transforms the concept into an image, but in such a way that the concept remains within the bounds of the image, can be grasped and possessed and, finally, expressed. In the symbolic system, the idea remains forever active and beyond the reach of the image; it remains thus inexpressible. An Aristotelian tradition treats the image as a form of didactic discourse; the image both supplies the absent meaning, taking its place, and bestows new meaning. And it is this theatre, where idea and discourse, iconic space and dramatic poem, coincide, which I propose to analyze here.

The poet is an image-maker; his discourse is necessarily a figurative one, and thus an indirect form of knowledge. We shall see how the sun allegory of 1661, a form both transparent and secret, whose poetic form has didactic significance, is stripped of meaning in the 1683 version. This study will concentrate mainly on the scenography and staging of the prologue and Act V in each version. But before addressing these questions, I would like to digress briefly to define what we mean by sustained metaphor, prologue and myth.

ON SUSTAINED METAPHOR . . .

The sun allegories serve to unmask the underlying discourse. Their role is to conserve and protect, for they are based on an underlying intertextual code ingrained in us through generations of tradition. *Memoria* is, after all, the fourth element of Rhetoric, and these allegories are thus already inscribed in that great world picture, the *theatrum mundi*. Their role is also an innovative and progressive one, for they offer a locus for reciprocal exchange, on the paradigmatic level, between factual knowledge, which circulates within the discursive space of the prologue, and cultural knowledge, within the space of the mythological fable. Their meaning thus has both a static and a dynamic aspect.

The prologue is at least implicitly recognized as a space where the readable is made visible – a visibility that then requires that one "read" it; this process conjures up by allusion a mental image from the external world (a historical king). According to Furetière, a representation is an "image which calls to mind and memory an absent object, and depicts it as it is. When we go to pay our respects to a dead prince lying in state, we see only his effigy." To represent is thus to substitute something present for what is absent.[10] Representation, the sign of some thing, eventually grants presence to this absent thing, by affirming its existence through repetition.

Myth can only ever be a collective creation, that is, a narrative outline whose earlier meanings are constantly being called into question. Appearing as they do at the end of the play, Corneille's sun allegories constitute a "figurative" representation and a distilled form of the narrative fable that has just been performed; by reflecting back on the fable embedded within them, these allegories undergo the metaphorical process which allows them to produce a new, hyperbolic text. By this process, the captivating images of the sun actually reduce the narrative capacity of the myth by virtue of the very fact that the image is there (the equivalent of an act of condensation), and through the superimposition of images (the equivalent of an act of transference). Thus, the mythological Sun is reduced, he loses his almost infinite referential dimension in favour of a single character, a single function; the idealized figure has now become model ("Que je laisse un modèle aux rois à naître" – [That I may leave a model for kings as yet unborn]). The designated models both exclude other potentialities (which are precluded by the presence of this reductive image) and distort the myth by the very expansion of possibilities (the build-up of meanings as the discourse slips from the prologue to the fable); once made real, they re-describe another reality.[11]

Nevertheless, the figure must necessarily resemble, more or less faithfully, what it is intended to represent. It is supposed to reproduce in the visible world the "idea" expressed by the poet. And this is precisely what the playwright had in mind when he staged the Sun in the *Toison d'or*.

THE ROYAL ALLEGORY OF 1661

The prologue, the play in miniature added on to the fable, is an innovation on Corneille's part – "another kind of prologue," as the poet emphasises in his *Discours de l'utilité et des parties du poème dramatique*. Forming as it does a discourse-cum-programme of princely conduct, the prologue was presented as a sort of "Miroir des Princes,"[12] in which word and look are subtly combined, designed to evoke in the sovereign's mind the practice of various high moral virtues[13] which would serve him as a code of conduct, as did the numerous political treatises of the first half of the seventeenth century – treatises on the education of a prince, reflections on Machiavellianism or on the fundamental laws of monarchy. The political literature of the time, Balzac's *Le Prince* or Richelieu's *Testament politique*, for example, had already forged an image of the monarch.[14] And Corneille's precepts, expressed within a historical, political discourse with a pedagogical intent, served to establish the king on the throne. The ritual of *la gloire*, in accordance with the existing tradition of a politics of personal prestige to which Corneille adhered, underlined the "personal" paradigms of the Prince: the sovereign could see – the eyes are the signs of intelligence, reason and judgement – so he understood the misery of his people; he was magnanimous;[15] he was the "father" – the protector – of his subjects. The ruined landscape of the setting in which the prologue opens, a material representation of an apocalyptic universe, reflects a state of political disorder; the king's accession to power promises a time of renewal. Public rejoicing is echoed in the theatre by the transformation of the garden of the first act into a *locus amoenus*. The invented landscape takes on new importance as it becomes the receptacle of meaning, projected onto a physical space. Hostilities with Spain had come to an end, promising peace with the outside world, and the *Fronde*, which had united against Mazarin all those threatened by the transfer to an absolute monarchy, was over, bringing peace at home.

This prologue is a truly symbolic passage, a "vestibule éclatant,"[16] a "dazzling entrance-way" not unlike the triumphal arches erected for royal entrances, symbolizing the taking of a city. It is both pre-text to the fable, and political pretext – the circumstances of its time gave it a realistic dimension. Right from the prologue's opening lines, France, the personification of the geographical location, defines the space by giving it a real, historical referent, even while there is no possible confusion with the actual country. And although the image of the gods removes the action from real time and space and places it in the realm of myth, it is the prologue which provides the essential key to the events and transforms them into representation. The allegories (Victory, Peace and France) allow the world of the gods and the individual story of the king to intertwine; these allegories translate a certain mentality – the hopes and feelings of the poet – thus taking on a prophetic character which can now be seen unfolding in the present.

The king is not physically present, but the prologue is illuminated by the traces of his royal splendour; his majesty shines through, even though the person of the monarch is subject to an almost religious taboo. Hymen wears the Queen's portrait painted on her shield; positioned as it is (a stage effect, essentially, due to its mesmerizing power), placed high above other, lesser mortals, the image endows the princess with sacred qualities, exalts the power of love, and, by association, evokes the real monarch. The fragmentation of the royal subject, his dispersal in time and space, allows for the reconstitution of a kingly figure. The very historicity of power – the concept of taking the throne,[17] which will meld with the concept of the sun – is what authorizes and founds that power.

In Act V, the Sun, literal image of a cosmic force and royal *topos* (in which we find the destiny of the king foretold) bursts forth from his palace, blazing with light:

> Et déjà le Soleil pour nous prêter secours,
> Fait ouvrir son palais, et détourner son cours.
> (V, iv, 2145–46)[a]

He casts his bright rays over Jupiter's magnificence and Mars's dark forest.[18] Corneille attempts to outdo the painters in his depiction of the palace of the Sun, placed on the median axis at the very centre of the cosmos, a reflection of the heliocentric universe of the Renaissance. His palace is not unlike that of Fame in Ovid's *Metamorphoses*.[19] This display of contrasts presents three scenes simultaneously, reminiscent of the earthly stage set laid out in similarly-tiered rows and hierarchical arrangement; a three-part set inherited from the Middle Ages. On his appointed course from the other side of the planet at the beginning of the tragedy, the Sun now shows himself, his splendour unveiled, driving his chariot "tout brillant de lumière." Refusing to give in to the demands of his son Aætes, who pleads with him:

> Donne-moi tes chevaux à conduire à ta place;
> Prête-moi de tes feux l'éclat étincelant
> (V, eii, 2156–57)[b]

he nevertheless agrees to change his course. He rises up towards Jupiter in all his glory, his rays only increasing the other's splendour. As he greets Jupiter, the Sun realizes that he must continue his circular course and take his beneficent light elsewhere. As an intermediary between an earthly king – his son Aætes – and the god of gods, he is second to none but Jupiter, and the laws he is bound to interpret. He is subject to the common laws and to nature, just as the monarch was both *legibus solutus* and *ratione alligatus*: above the laws in principle, but bound by reason to respect them. Then the Sun sinks in the sky as if to disappear into the sea.

In bringing to the fore the idealized figure of the Prince in the final and all-embracing imagery of the sun, Corneille's precepts serve to enthrone the subject of this fulsome praise by focussing admiration on "sa conduite adorable", his beloved ways. The playwright believed in this ideal of excellence, virtue and goodness, a poetic, though not utopian, vision of the prince. The ideal prince was not a dream figure: everything seemed possible in 1661. The poet was thus conforming to the conventions of epideictic literature: the usefulness of the panegyric lay in its capacity to draw attention to the "person" of the man being so honoured, thus obliging him to conduct himself virtuously so as to resemble the model more closely. The benefits did not flow in one direction only, for a politically educated populace helped secure social stability. A society was constituted according to the natural order of the family, and ruled by the political order of the State, which the sovereign, schooled in various moral principles, embodied. Naudé, in his *Bibliographie politique* of 1642, emphasized the useful function of these panegyrics, which served both to persuade and command.[20]

This representation was part of the constructed fiction of the monarchy, based on an ideology that both manipulated political writing and political practices, and that lay at their very heart. The portrait of the king, destined for Louis XIV, was not that of a real monarch, but rather that of the potential ruler just beginning his reign. This is what Corneille recalls in his *Remerciement au Roi* of 1663 (the emphasis is mine):

> Afin d'ouvrir ma scène encore un coup pour toi:
> J'y mets la Toison d'or [. . .]
> On te voyait dès lors à toi seul comparable
> Faire éclater partout ta conduite adorable,
> Remplir les bons d'amour, et les méchants d'effroi.
> Jusque-là toutefois *tout n'était pas à toi*, [. . .]
> Maintenant qu'on te voit en digne potentat
> Réunir en ta main les rênes de l'État,
> *Que tu gouvernes seul*, et que par ta prudence
> Tu rappelles des rois l'auguste indépendance,
> Il est temps que d'un air encore plus élevé
> Je peigne en ta personne un monarque achevé;
> Que je laisse *un modèle aux rois* qu'on verra naître
> Et qu'en toi pour régner je leur présente un maître.[c]

THE PROLOGUE OF 1683: ROYAL MIRROR IMAGES

Within the identical framework-settings of the prologue[21] and Act V[22] – a metaphor for a closed and immutable society – both the Sun and the portrait of King Louis XIV follow a similar path. Emerging from a magic mirror in the prologue, the portrait with its univocal signified, borne, as though by three Graces, by Mercury, Fame and Fortune (allegorical figures in vogue at the time), receives homage from all the characters present and takes wing under escort for some mythical other world no longer symbolized by France.

The king's face, the only realistic element, is completely overshadowed by the painting that frames it. Indeed, the portrait of the kind falls within the framework of various allegorized virtues and a whole constellation of symbols represented by divinities. The virtues of appearance and ceremonial display take precedence over the intellectual virtues, but it is the Arts which educate the monarch and immortalize his image. Jupiter allows all the gods to offer their services, then shows them an "assemblage" befitting an anamorphosis with mirrors:

> Je vais dans un seul ouvrage
> De toutes vos vertus montrer un assemblage.
> (Prologue, 3)[d]

This syncretic portrait of the king is the model that Jupiter proposes to the dauphin:

> Voilà le seul objet de grandeur et de gloire
> Que du jeune Héros doit remplir la mémoire.
> (Prologue, 3)[e]

Jupiter invites all the characters to come and admire this portrait which holds everyone's gaze, and becomes the protagonist itself through a reversal of roles – the actants become beholders [*regardants*] and take part in the discourse. Thus the narrative is also embedded within the framing narrative; the king's effigy stands out as though on a coat of arms, and the discourse becomes a heraldic device. This heraldic representation was intended for the king who would be watching himself, just as an informed public would be admiring him.

The daystar of the last act[23] occupies the central position between the palaces of Jupiter and Mars, spanning the entire universe in his rotations, then addresses Jupiter and the public as he rises slightly. His revolutions have taken him round his heavens, where he stops for a moment to retrace his steps. He begins to grow, rapidly doubling in size. As he starts to dominate the stage, the sun figure takes on tremendous presence; he bathes everyone in his light, holding the audience in his thrall. The author focusses all the spectators' (or reader's) attention on this hugely enlarged and imposing image, a reproduction of Louis XIV's portrait, in which they lose themselves. The Sun, an extraordinary device for the celebration of self, allows the king to see himself presented, and re-presented. Thus removed from the temporality of the dramatic discourse, he enters the sphere of the Eternal. The historical dimension shrinks, then vanishes: time becomes the time of the kind, just as at Versailles the present became eternity. Twenty years of absolute monarchy separate the two sun figures – glorious military campaigns and the establishment of the Academies put all forms of expression at Louis XIV's disposal.[24] It was the zenith of absolutism, and all political debate had come to a close.[25]

Picture within picture, performance within performance, narrative within narrative, play and re-play – these are the devices La Chapelle used to pay homage to the king in his *Desseins*; the poet was conforming to long-established, and restrictive, conventions. He offered up an embellished, but simplified, image of Louis XIV, and we might say, as did Père Le Moyne in 1666 in his *Épitre* to *De l'Art de régner*:

The very amusements which your Majesty sometimes engages in are always most noble and royal. . . . One might even say, Sire, that you are the most beautiful and most rare show piece in these entertainments; the richest and most unusual decoration of the spectacles to be seen; that you are a spectacle in your own right who catches everyone's eye and draws the spectators away from all else.[26]

The king, on stage, becomes both the subject and the *destinataire* of the spectacle, to whom the action is directed, as in court ballets in which the king was the actor and embodied two personalities, one real and the other fictitious.[27] Père Le Moyne went on: "These characteristics, Sire, which do not consist of outward adornments or precious stones, but which are the mark of God's finger upon you, and are inherent to your Person, can never be taken from you." This civic and spiritual bipolarity lends a certain ambiguity to the allegory;[28] this is not a new phenomenon, since the monarch was the king by divine right, but rather constitutes an extension of the old mediaeval doctrine of the two bodies of the kind. The painted Sun was thus certainly the symbol of the Prince, the image of God, who conferred upon the religious conception of the royal function a new intensity: in the chapel of Versailles the faithful turned their eyes towards the king who, alone, gazed upon the altar: for the king was God.

FROM THE "ENIGMATIC" SUN ALLEGORY OF 1661 TO THE CONVENTIONAL ALLEGORY OF 1683

In 1661, the imaginary body of the king – a sun *topos* – was the incarnation of the monarchy, the basis of a spectacle whose goal was to consolidate the State. It had a political function; and educative one – to point the way towards wise kingship; a moral function – for the king wished his people to love him. Corneille was evidently doubly anxious to persuade when he put forth his model man of State. Jurieu reminds us of this in his *Soupirs de la France esclave* of 1691: "In the past we talked of nothing but the interests of the State, the needs of the State. Now to talk this way would be, quite literally, a crime of *lèse-majesté*. The King has taken the place of the State." And in fact Louis XIV had, in taking power, appropriated all the signs of the exercise of power, thus leaving no room for others to use them or to signify the State on his behalf.

La Chapelle's sun figure is fixed in time, he has almost brought life to a standstill. On the sidelines of history, the figure looks out at a public dazzled by his own radiance. He has been somehow "*versaillisé*": made permanent, unmoving, immortalized in the present moment. The portrait of the king in the prologue becomes grafted onto the sun image, thereby solidifying and completely realizing the allegory. The two terms of the sustained metaphor move within an "internal" space, between the sign (the king's portrait) and the object (the Sun), without considering the poetic "external" space. The distinction between the portrait of the king, who is not the real king, and the Sun is diminished in the version of 1683. This representation of the sun is a metaphor *in praesentia*.[29] And the tension between the similarities and differences of the two terms has now been eliminated; now the two no longer simply resemble one another, they are identical. As Père Bouhours wrote in the *Entretiens d'Ariste et Eugène*: "Since the king adopted the sun as his symbol, and appropriated this splendid star for himself, if I may say so, enlightened people mistake the sun for the monarch; they conceive the two things simultaneously."[30]

The sun figure depicted by La Chapelle for the Duc de Bourgogne represents a model of the prince with Louis XIV's features, a form of the truth.[31] The sun metaphor, a figure now fixed and finite in meaning, is drained of substance and returns to nothingness; and thus through over-use, it returns to its symbolic status. The imaginary body of the kind of 1683 melds together the image of Louis XIV and that of the Prince; he has a primarily celebratory function: to win the admiration of his subjects. The poet's aim was to create a compelling image of the monarch, and not to reflect on what it means to be a king. La Chapelle's image bears the King's features and is thus reduced to a king who personifies the State. Corneille's figure is diffuse, dispersed within the discursive space.

The poet paints a more subtle portrait in the 1661 version, a composite picture of the ideal prince whom the king is supposed to emulate, and which reconciles the real and the imaginary, enhancing the real with the heightened colours of the desire of the collective psyche. It is a metaphor *in absentia*,[32] a *living* metaphor which is both dynamic and enigmatic.[33] The fable, itself a sustained metaphor (by Furetière's definition), endows this princely metaphor with a rhetorical and poetic function. And indeed, on the paradigmatic level, the signs of royalty present in the prologue gradually merge into the mythological figure of Jason;[34] then, through a process of reflection and refraction, they are taken up by the sun imagery (the sun myth is in a way sapped of its meaning by the model image, yet reinvested with meaning by this same image, through allusion). Thus, allegory appears as "a mooring-point in the ideological struggle,"[35] where the political aspect enters in. By 1683, however, the historical and political discourses have given way completely, in favour of the allegorical discourse alone. It would seem that La Chapelle was indeed bound by the conventions of his time.

By depleting the sun allegory in this way, the mythological motif of 1683 is reduced to mere ornament: any political interpretation, from prologue to exodium, is now impossible within the confines of the discourse. Corneille was already aware of this reduction of the fable when in 1667 he wrote in *Au Roi, sur son retour de Flandres*: "J'y porte, au lieu de toi, ce héros dont la gloire/ Semble épuiser la fable et confondre l'histoire."[f] And even Jason meets his match in this man whose towering intellect is "supérieur" to his own: "Recalling the fleece of the Fable and our own Great Monarch's latest conquests in Flanders and Burgundy, we could say without flattery that His Majesty the conqueror is even greater than Jason."[36] "Greater", perhaps, but mesmerized by his own personal glory. The myth has become no more than an occasion for entertainment, no longer serving as an exemplary and signifying narrative.

Corneille and La Chapelle are making "mythistory" in their *Toison d'or*; they are translating into images the imaginary body of the king (according to humanist rhetoric), using the mythological and historical references that sustained the monarchy. These figures, founded on the glory won by battle, wore the face of Victory in 1661 and that of Fame ("la Renommée") in 1683, banishing any idea of defeat, the antithesis of victory and glory. Twenty years after the first *Toison d'or*, absolute monarchy reigned supreme. Fame, loudly proclaiming her victories, now had Fortune by her side. Sun allegories did not represent or evoke a particular reigning monarch with individual features, but an ideal Prince in the 1661 version (without Roman mask) and a mythical Prince by 1683. The king was the prisoner of the image he was given. In 1683 he put himself on the stage; his power was now lapsing into the realm of the theatrical – of scenery, stage effects and allegories – which staged his brilliance iconically. For the king was not king by nature; he was forced to play the role – become representation, in other words – in order to be king in reality. Absolutism had thus become an act of the imagination.

Translated from the French by Marianne Krajicek, University of Alberta.

Université Concordia
Montréal

NOTES

[1] *Mercure galant*, 1684, pp. 3–9, quoted in G. Sabatier, "Le Parti figuratif à Versailles," *XVIIe Siècle* (1988), p. 419.

[2] For Cappel's speech, see A. N. P 2306 [Le Maistre], fols. 353v-369v (15th January 1537); quoted in Sarah Hanley, *The* Lit de Justice *of the Kings of France* (Princeton, New Jersey: Princeton University Press, 1983), p. 90.

[3] Dupuy, *Traité de la majorité* [Reg. du Parl.], pp. 467–69; quoted in S. Hanley, *ibid.*, p. 231.

[4] Louis XIII was named the "petit phénix" and *Sol Franciae* in 1611; see S. Hanley, *ibid.*, p. 264.

[5] "The theme of a Pythian Apollo [. . .] was used to glorify the triumphant victory of the monarch over the pernicious forces of the opposition" – F. Bardon, *Le Portrait mythologique à la Cour de France sous Henri IV et Louis XIII* (Paris: Éd. Picard, 1974). This is a recurrent theme in royal imagery.

[6] Louis XIV, *Mémoires et divers écrits* (Paris: Club du Livre, 1960), pp. 124–25.

[7] N. Ferrier-Caverivière, *L'Image de Louis XIV dans la littérature française de 1660 à 1715* (Paris: PUF, 1981), p. 76.

[8] Le Moyne, *De l'Art de régner* (Paris, 1665), p. 45.

[9] J. Jehasse, "Guez de Balzac et Corneille face au mythe romain," in *Pierre Corneille, Actes du colloque tenu à Rouen du 2 au 6 octobre 1984* (Paris: PUF, 1985), p. 249.

[10] L. Marin, *Le Portrait du roi* (Paris: Éditions de Minuit, 1981), pp. 9 and 10. Marin points out that "représenter" means both "to present again" (in the time sense) and "in the place of", speaking spatially.

[11] P. Ricœur, *La Métaphore vive* (Paris: Seuil, 1975), p. 11.

[12] Ronsard, quoted in T. Murray, "Richelieu's Theater: The Mirror of a Prince," *Renaissance Drama* 1977, p. 276. Ronsard wrote in *Fin pour une comédie*: "La Comédie apparaît un exemple/ Où chacun de son fait les actions contemple:/ Le monde est le théâtre, et les hommes acteurs."

[13] The first definition that Furetière gives of virtue is "the power of action inherent in all natural bodies, according to their qualities or properties". Corneille uses the famous mediaeval square of the cardinal virtues: justice, prudence, temperance and courage.

[14] A. Stegmann itemizes the political treatises published in France from 1600 to 1648 under several headings: sociology of the authors, from first principles to the mechanism of institutions, conflict of values, dialectics of values, in "Les Traités politiques en France (1600–1648): entre le juridique et le moral," *Cahiers de Littérature du XVIIe siècle*,

9 (Toulouse, 1987), pp. 77–109. For the education of Louis XIV, please see the following works: Gomberville's *Discours des mœurs*, Vauquelin des Yvetaux's *Instruction du prince*, Arnauld d'Andilly's *Mémoires pour un souverain*, and the various treatises by La Mothe le Vayer. All these texts deal exclusively with the greatness of the king.

15 The virtues of magnanimity and of magnificence are inseparable and are synonymous with the French virtue of generosity. See M. Fumaroli in the discussion of J. Rohou's paper, "La Fronde et la vision de l'homme de la 'générosité' à l'avidité" in *La Fronde en question, Actes du 18e colloque du CMR* **17** (Aix-en-Provence, 1988), p. 381.

16 Marmontel, *Éléments de littérature* IV, p. 154, quoted in C. Girdlestone, *La Tragédie en musique (1673–1750) considérée comme genre littéraire* (Paris: Droz, 1972).

17 It is worth noting that it was in July 1527 that the registers of the Parlement first call the *Lit de justice* the "royal throne". The word "throne" had until that time meant not the royal seat itself but the Grand' Chambre as a whole, or indeed the larger spatial complex where the king and Parlement met. See S. Hanley, *op. cit.*, p. 61.

18 Here is the description of the palace of the Sun given in the stage directions for Act V in the 1661 version:

"Le ciel s'ouvre, et fait paraître le palais du Soleil, où on le voit dans son char tout brillant de lumière s'avancer vers les spectateurs, et sortant de ce palais, s'élever en haut pour parler à Jupiter, dont le palais s'ouvre aussi quelques moments après. Ce maître des Dieux y paraît sur son trône, avec Junon à son côté. Ces trois théâtres, qu'on voit tout à la fois, font un spectacle tout à fait agréable et majestueux. La sombre verdure de la forêt épaisse, qui occupe le premier, relève d'autant plus la clarté des deux autres, par l'opposition de ses ombres. Le palais du Soleil, qui fait le second, a ses colonnes toutes d'oripeau, et son lambris doré, avec divers grands feuillages à l'arabesque. Le rejaillissement des lumières qui portent sur ces dorures produit un jour merveilleux, qu'augmente celui qui sort du trône de Jupiter, qui n'a pas moins d'ornement."[8]

19 Ovid, *Metamorphoses*, Book XII, 11. 39–45: "There is a place in the middle of the world, 'twixt land and sea and sky, the meeting-point of the three-fold universe. From this place, whatever is, however far away, is seen, and every word penetrates to these hollow ears" – Frank Justus Miller, translator (London: William Heinemann Ltd, 1984).

20 "Panegyrics are most important, and are of no small use in the instruction of kings, especially given that so many illustrious examples, so many heroic, almost divine, virtues are becoming common knowledge, and are easily understood when they are conveyed by the speeches of these great orators."

21 "*La Toison d'or*, tragédie en machines de M. Corneille l'aîné, représentée sur le théâtre royal des seuls comédiens du roi entretenus par Sa Majesté en leur hôtel, rue de Guénégaud. Avec un prologue nouveau. Entreprise sous la conduite du sieur Du Fort, Paris, impr. de Vve G. Adam, 1683, 35 p."

The *Desseins* of 1683 read as follows:

"Tandis qu'Aæte se plaint de la perfidie de sa fille, Junon paraît dans son Char pour remonter au Ciel, presque en même temps le Ciel s'ouvre et fait paraître le palais du Soleil au-dessus du Phase. Le Soleil paraît tout éclatant de lumière et s'avance du côté des spectateurs, en s'élevant insensiblement comme pour parler à Jupiter, dont le palais au-dessus de ce Soleil s'ouvre aussi quelques moments après de sorte que ces deux

Palais tout brillants se font voir au-dessus de la forêt de Mars, et forment trois théâtres qui font un spectacle imitant le mouvement des Planètes, après s'être avancé vers les spectateurs, s'élève et traverse le palais de Jupiter, au-dessus duquel il s'arrête un moment, revient ensuite dans sa première place où il semble s'augmenter, et s'agrandissant tout d'un coup, il se fait voir une fois plus large qu'il n'était; il disparaît ensuite en s'abaissant pour fondre dans la mer, et celui de Jupiter en remontant dans le ciel finit tout ce grand spectacle."[h]

22 The setting for the Prologue of 1683:

"Une manière d'amphithéâtre d'architecture rustique se présente à la vue, plusieurs terrasses élevées les unes sur les autres, et qui se communiquent par des escaliers qui sont des deux côtés du théâtre forment la décoration du Prologue. La première terrasse est bordée des deux côtés par des grands bâtiments de marbre blanc soutenus par des chênes de pierre, entre lesquels sont placées dans des niches quatre figures représentant les quatre Vertus principales. Cette terrasse est bordée par une balustrade en raccord qui lui sert d'appui, sous la balustrade sont des niches dans lesquelles deux fontaines font des jets bouillonnants, qui forment ensuite de grosses nappes. Sur le haut de cette balustrade sont aux deux côtés de grandes allées de cyprès et au milieu l'escalier qui monte à la seconde, aux deux côtés de laquelle deux escaliers pour monter à une troisième couverte d'arbres qui bornent agréablement la vue, forment un fer à cheval orné au milieu de plusieurs figures dans les niches. C'est sur cet amphithéâtre que la France, Jupiter et tous les autres dieux paraissent."[i]

23 The setting for Act V of the 1683 version:

"La forêt de Mars composée de toutes sortes d'arbres extrêmement touffus, auxquels sont attachés des trophées d'armes, est la dernière décoration qui présente le merveilleux spectacle de la Toison d'or. Au milieu de cette forêt paraît un arbre plus élevé que les autres auquel la Toison est attachée; dans l'éloignement du théâtre, on voit la rivière du Phase avec le navire Argo, qui semble n'attendre plus que Jason et sa conquête pour partir. Lorsque Jason se prépare à l'aller enlever, Médée paraît sur un dragon si bien représenté, dont les mouvements sont si beaux, si justes et si naturels que les yeux sont trompés, et que le spectateur ne saurait regarder ce monstre qu'avec une espèce d'effroi. Elle prend la Toison et la met sur le col du dragon; Zéthes et Calaïs, Argonautes ailés, paraissent et s'élèvent au plus haut des nuages en croisant leur vol. Il se forme entre eux et Médée, qui pare les coups qu'ils portent au dragon, un combat merveilleux. Enfin ils s'envolent comme vaincus, et Médée les suit sur le dragon avec la Toison qu'elle va porter à Jason. Tandis qu'Aaete se plaint de la perfidie de sa fille, Junon paraît dans son char pour remonter au ciel, presque en même temps le ciel s'ouvre et fait paraître le palais du Soleil au-dessus du Phase."[j]

24 The *Gazette de France*, the *Mercure galant* and the *Journal des Savants* all report the military exploits of Louis XIV. In 1663 the king founded the Petite Académie (for the study of the arts and humanities), in 1662 the Academy of Painting and Sculpture, in 1671 the Academy of Architecture, followed by the Royal Academy of Sciences, and the Royal Academy of Music. In 1672, after Séguier's death, he became the patron of the Académie française.

25 G. Ferreyrolles, *Pascal et la raison du politique* (Paris: PUF, 1984), p. 18: "It has

become impossible to express any form of criticism which does not attack the king directly: the minister is no longer there to serve as a scapegoat, and it would be unthinkable to find fault with the Prince. In matters of state, we shall have to settle for philosophizing on the education of the dauphin."

[26] Le Moyne, *De l'Art de régner* (Paris, 1665), p. 46.

[27] Louis Marin's remarks support Père Le Moyne's reflections: "The Court has become the great new theatre: the whole political arena has been transformed by this aesthetic shift towards the theatrical. The Court at home and the war abroad now constitute the new scenes of action. For the monarch in the 'theatre-come-true' of his Court, there can be no theatre of politics [*la politique*], and even less of political topics [*le politique*], since politics and the political side of things have themselves become theatre," in "Théâtralité et politique au XVIIe siècle: sur trois textes de Corneille," *La France et l'Italie au temps de Mazarin*, texts selected and published by Jean Serroy (Presses universitaires de Grenoble, 1986), p. 404.

[28] Ernst Kantorowicz, *The King's Two Bodies: A Study in Medieval Political Theology* (Princeton, New Jersey: Princeton University Press, 1957). The ambiguity expressed by the maxim *le roi ne meurt jamais* (the king never dies) tended to blur the distinction between the king's two bodies (the person and his function).

[29] Paul Ricœur, *La Métaphore vive*, p. 209: in metaphor *in praesentia*, "the two terms are present together, as well as the mark of their partial identity".

[30] Bouhours, "Les Devises, sixième entretien," *Les Entretiens d'Ariste et d'Eugène* (Paris: Bibliothèque de Cluny, 1962, reprint of first edition of 1671), p. 178.

[31] Nietzsche, *Le Livre du philosophe/Das Philosophenbuch*, p. 183; quoted in C. Normand, *Métaphore et concept* (Paris: PUF (Dialectiques), 1976), p. 24: according to Nietzsche, "truths are illusions whose illusory nature we have forgotten, metaphors which have worn through and lost their sensual power, coins on which the effigy has been rubbed away and which are now only of interest as pieces of metal".

[32] Ricœur, p. 209: in metaphor *in absentia* ("true metaphor according to the Ancients"), "the substitutable term is absent from the discourse".

[33] The Greek word *aenigma* "certainly evokes the idea of obscurity, but also the even more intangible idea of a representation of an image, indeed of a narrative" – J.-F. Groulier, "Monde symbolique et crise de la figure hiéroglyphique dans l'oeuvre du Père Ménestrier," *XVIIe siècle* **158**, 1988, p. 101.

[34] Ménestrier, *L'Art des emblèmes* (Paris: La Caille, 1684), p. 33. *Éloges et Discours sur la Triomphante réception du Roi en sa ville de Paris* . . . , A Paris, chez Pierre Récolet, 1629 (Arsenal Ra4 233), page 77 reads: "That very ship is France whose people are the most war-like living today: her captains and soldiers are so many Heroes and Argonauts: but the King is her Jason, who by his valour and wisdom, like another Minerva, stands at the helm of this great monarchy, guiding her to the contentment of peace, and to the fatal fleece which is the love of his people and the union of his subjects." The king so described is undoubtedly Louis XIII.

[35] C. Normand, *op. cit.*, p. 8.

[36] Abbé O. F. de Bianville, *Symbole héroïque pour les dernières conquêtes du roi* (Paris, 1688); quoted in N. Ferrier-Caverivière, *op. cit.*, p. 100.

PROSE RENDERINGS OF CORNEILLE

[a] And now the Sun comes to our aid/ Opening his palace and turning from his heavenly course

[b] Let me drive your horses in your place./ Grant me a spark from your own fiery radiance

[c] To inaugurate my play, I offer one more glory in your name:/ I give you the Golden Fleece [. . .]/ You were supreme even then, without compare;/ Far and wide, your beloved ways were known,/ Filling good men's hearts with love, striking terror among the wicked./ Until this day, however, *the world was not your own* [. . .]/ Now we see you master of your royal destiny/ With the reins of State firmly in your hands,/ *In your hands alone*, and your wisdom/ Recalls the regal independence of a true king,/ It is meet that I paint you in even loftier terms,/ In a portrait fit for a consummate king;/ That I may leave a *model for kings* as yet unborn,/ And that in my play they may find the veritable master they seek.

[d] I shall show you a work/ Wherein all your virtues are united in a single stroke.

[e] Here is the only object of grandeur and glory/ On which the young Hero should set his sights.

[f] I bring to this play, in your place, that hero whose crowning glory/ Seems to beggar the fable, and leave history dumbfounded.

[g] The heavens open and show the palace of the Sun, where he is seen in his chariot, blazing with light, travelling towards the audience, and as he leaves his palace he rises up to speak with Jupiter, whose palace also opens a few moments later. This master of the Gods appears there on his throne, with Juno at his side. These three stage sets, all visible simultaneously, create a most pleasing and magnificent spectacle. The dark foliage of the thick forest, which occupies the first theatre, sets off the brilliance of the others all the more clearly, with its contrasting shadows. The palace of the Sun, which comprises the second theatre, has columns of glittering shreds of cloth and gilded panelling, with great swathes of ornamental foliage. The play of light from this gilding produces a marvellous effect, intensified by the glitter of Jupiter's throne which is no less lavishly embellished.

[h] *La Toison d'or*, tragedy with stage effects by Monsieur Corneille the elder, performed in the royal theatre by the king's own actors entertained by His Majesty in their private residence, Rue de Guénégaud. With a new prologue. Under the management of the *Sieur* Du Fort, Paris, at the printing house of Widow G. Adam, 1683, 35 pp.

The *Desseins* of 1683 read as follows:

"As Aætes complains of his daughter's perfidy, Juno appears in her Chariot and ascends into the Heavens, and at the same moment the Heavens open to show the palace of the Sun over the River Phasis. The Sun radiates light and moves towards the audience, rising gradually as though to speak to Jupiter, whose palace above this Sun also opens a few moments later, so that the two glittering Palaces are displayed above the forest of Mars, forming three sets which create a marvellous spectacle imitating the movement of the Planets. The Sun, after journeying towards the audience, rises up and crosses Jupiter's palace, above which he stops for a moment, then returns to his original position where he seems to grow ever larger, doubling his size. Then he disappears, sinking down as though into the sea, and Jupiter concludes the display, as his palace rises into the heavens."

[i] A sort of amphitheatre in rustic style is seen, several terraces placed one above the other, linked by staircases on both sides of the theatre, forming the decoration for the Prologue. The first terrace is lined on both sides by great white marble buildings supported by stone oak trees, between which four figures, representing the four principal Virtues, stand in alcoves. This terrace is bordered by an adjoining balustrade which supports it, and under the balustrade two fountains play, throwing up great cascading jets of water. On top of this balustrade are great avenues of cypresses on either side, and in the centre is the staircase which goes up to the next level, bordered in its turn by two staircases leading to a third level covered with a charming screen of trees, and forming a horse-shoe shape decorated in its centre with several figures in alcoves. France, Jupiter and all the other gods are to appear in this amphitheatre.

[j] The forest of Mars, made up of all sorts of very densely wooded trees, to which are attached weapon trophies, is the last set to be seen in the wonderful spectacle of the *Toison d'or*. In the middle of this forest stands a tree, higher than the others, to which the Fleece is fastened. In the distance, the River Phasis can be seen, with the ship *Argo* apparently waiting only for Jason to complete his exploit before setting sail. While Jason is preparing to go and claim the Fleece, Medea appears on a dragon so lifelike, whose movements are so accurate and so natural, that the spectator cannot gaze upon this monster without terror. Medea takes the Fleece and places it round the dragon's neck; Zetes and Calais, winged Argonauts, appear and fly up to the highest cloud, circling round one another. The two of them stage a spectacular battle against Medea, who staves off their attacks on the dragon. Finally they fly off, defeated, and Medea rides off after them on the dragon, carrying the Fleece which she will take to Jason. As Aætes complains of his daughter's perfidy, Juno appears in her chariot and ascends into the Heavens, and at the same moment the Heavens open to show the palace of the Sun over the river Phasis.

FRANÇOISE DIVORNE

LE JEU ALLÉGORIQUE CHEZ POUSSIN: *Les* Putti *comme emblème dans* Tancrède et Herminie *(Birmingham)*

Je souhaite montrer que dans ce tableau de Nicolas Poussin, les *putti* sont emblème, esprit du *concetto* qui permet au peintre de matérialiser de l'invisible avec du visible. La figure de l'allégorie est ce lien mouvant, processus de mise en relation de deux mondes qui permettra, ainsi que l'écrit Giordano Bruno dans le *De umbris idearum*, que "tant que celui qui regarde a les yeux absorbés par l'image, il ne soit pas envahi par l'amour, mais dès que l'esprit s'arrachant à l'image rentre en soi-même et conçoit l'au-delà du visible, surgisse l'amour". Ces *putti*, figures de l'Eros antique, héritiers des *daïmones* pourraient renvoyer à ce noeud d'images et d'idées que partagent, dans l'imaginaire, le peintre et le spectateur.

A detail from "Tancred and Erminia" by Nicolas Poussin. From the Collection of The Barber Institute of Fine Arts, The University of Birmingham.

M. Kronegger and A-T. Tymieniecka (eds.), Analecta Husserliana XLII, pp. 113–130.

N'est-ce pas la manière de l'artiste, en effet, comme celle de l'Eros, d'unir toujours des choses diverses et opposées, de chercher dans le "visible" l'"invisible", dans le "sensible" l'intelligible.[1]

UN POINT THÉORIQUE SUR L'ALLÉGORIE

Dans l'usage actuel, les noms *symbole, allégorie, métaphore* sont distingués les uns des autres dans une sorte d'approximative nébuleuse, "les adjectifs "symbolique, allégorique, emblématique" sont devenus interchangeables".[2] C'est méconnaître les qualité propres de l'allégorie qui, hors de son usage banalisé et universel, est une véritable machination du sens qui *signifie par l'autre.*

L'allégorie que je souhaite défendre n'est pas la figure figée, "cette image banalisée de concepts", figure froide, compassée, rigide, trop obscure ou trop transparente, qui, par exemple, d'une figure de femme portant balance assène pompeusement, au XIXe siècle, I'idée de justice.

L'allégorie qui m'intéresse est "ce véhicule si propre à l'art qui a déjà par nature son origine dans l'art, très exactement dans la comparaison, et dont l'essence tient aux effets de ricochet, à travers des époques et des objets différents, d'un *eadem sed aliter, aliter sed eadem* soutenu par l'analogie; l'allégorie se fonde et ne cesse d'en appeler à des correspondances qui traversent de part en part le monde. L'allégorie cerne un sens au moyen d'une pluralité de significations et elle ne contient pour cette raison que des archétypes de l'éphémère, des archétypes dont la signification va toujours dans le sens d'une *alteritas* à la différence du symbole, qui demeure pour sa part constamment subordonné à l'unité d'un même sens".[3] Ernst Bloch situe l'allégorie dans une perspective utopique où le sens ultime du monde n'est jamais donné à l'avance mais en train de se faire et dont l'issue est incertaine. Chez Bloch, le cosmos est une multitude d'univers et non un univers, la réalité, plusieurs niveaux qui cohabitent en même temps, un espace ouvert qui n'a pas trouvé son *arkhé.*

Posant le rapport de l'homme au sens, l'allégorie est, dans une oscillation entre ce qu'elle montre et ce qu'elle voile, un instrument privilégié de la connaissance. Ce qu'elle montre avec le plus d'ostentation est ce qui est le plus indéchiffrable. Recélant un autre sens sous le sens, mixte de parole et de silence, elle voile ses secrets et suscite en nous la quête infinie d'un sens qui se dérobe.

Les auteurs latins, Quintilien le premier, présentent dans la rhétorique, la figure de l'allégorie dans son régime d'altérité. "Allegoria, quam inversionem interpretantur, aut aliud verbis, aliud sensu ostendit, aut etiam interim contrarium."[4] "Aliud verbis, aliud sensu" se prêtant à plusieurs interprétations: un sens autre que celui des mots, disant une chose et voulant en dire une autre, autre chose par les mots, autre chose par le sens.

L'allégorie possède une valeur instrumentale qui permet d'étudier une oeuvre, tableau ou texte, c'est-à-dire un objet en mouvement et dans cet objet le mouvement.

L'intérêt de l'allégorie tient dans sa propriété de "mise en relation" dont l'enjeu serait la signification d'une chose: une oeuvre (tableau ou texte), à propos d'une autre chose: d'une autre oeuvre – tableau ou texte; d'un texte par un tableau; d'un tableau par un texte, etc . . . Ce processus tient du fonctionnement du signe lui-même, avec la triple ambivalence que recèle ce terme chez Aristote.

Si par signe, on entend *seméion*, l'allégorie aura pour support essentiel le langage, les mots dans la peinture ou dans le texte (le *motto* en peinture, l'intertextualité pour le texte).

Si par signe, on entend plutôt *eikos*, l'allégorie aura pour support une oeuvre, texte ou peinture et renverra à une duplicité: le texte, l'image pourront être entendus, lus, selon le code herméneutique ou iconographique à partir d'une *alteritas* mais aussi de l'idée de dissimulation que contient l'étymologie de l'allégorie. En effet, le mot *hyponoia* désigne chez les poètes homériques, "l'enseignement théorique dont on "pense" qu'il est "sous" le revêtement imagé".[5] Plutarque dans un passage de *Sur la lecture des poètes*, enregistre la transition de *hyponoia* à *allegoria*. "Au moyen de ce que les Anciens appelaient des significations cachées et que l'on nomme aujourd'hui des allégories". *Allo-agoreuein*, c'est délivrer une parole autre, déclarer publiquement (devant l'*agora*) autre chose que ce qu'on dit; *hyponoia* suggère le sens cryptique, *allegoria*, l'altérité. Le pseudo-Héraclite dans le *De vita et poesi Homeri* affirme que le nom lui-même d'allégorie suffit presque, tant il correspond à en indiquer l'essence. "L'allégorie fait comprendre une chose par le moyen d'une autre".

Si par signe on entend *techmérion*, l'allégorie renverra sans ambiguité à l'idée de code pictural fort.

Mais il faut souligner la grande subtilité de l'allégorie qui mêle ces diverses catégories du signe telles que les pose Aristote dans *Rhétorique I et II.*

Métaphore continuée comme la définit Quintilien, donc se proposant de "dissimuler" à l'infini, l'allégorie entretient des rapports avec la métaphore, l'énigme qui l'obscurcit, sauf si ce rapport est justement dosé ainsi que le souligne saint Augustin dans *De Trinitate* dans un passage relatif à l'exégèse de *I Cor. XII*: "Videmus nunc per speculum in aenigmate".

Je souhaite faire comprendre que pour cerner l'essence d'une chose il faut s'en approcher au plus près sans y entrer avec crudité. La mise à nu dévoie le sens, aveugle le sens. Le sens voilé est le combat inverse du sens aveuglé.

Ce que je veux dire c'est qu'entre une oeuvre – la création d'une oeuvre –, et le regard qui sera porté sur cette oeuvre, l'allégorie est le véhicule qui, à l'image du medium véhiculant le pigment en peinture, instaure un "langage" entre la création et le regard.

Plutôt que de construire une grille conceptuelle systématique ou de passer les thèmes de mon travail comme aux rayons X, ce qui expliquerait peut-être tout sans par conséquent permettre à l'essence du sens d'advenir, je voudrais opérer un parcours entre les lieux qui relient oeuvre créée et regard porté sur cette oeuvre.

Je voudrais examiner, à partir des traits fondamentaux, structuraux de l'allégorie – le sens caché, l'*alteritas*, le mouvement – et de leur contextualisation généalogique renvoyant à des variations dans une même fonction, dans des motifs différents, la pertinence de cette allégorie en tant que processus de connaissance.

Ainsi, un tableau est le résultat du travail de plusieurs imaginaires, de divers "lieux de mémoire":

– *la "storia" dont s'inspire le peintre*
- elle peut être une histoire (mythe, etc . . .),
- elle peut s'ancrer sur "le motif",
- elle peut prendre sa source directement dans l'imaginaire du peintre, de l'écrivain, du sculpteur, etc . . .

Dans tous les cas, il en va d'une mise en jeu de l'imaginaire du créateur de l'oeuvre; de la *storia* d'un autre que garde-t-il, qu'en fait-il?

Que se passe-t-il quand Poussin peint *Tancrède et Herminie*, qu'invente-t-il? S'agit-il de *phantasia*?

Je souligne ici la différence entre *mimesis* et *phantasia* et rappelle dans la *Vie d'Apollonios de Tyane* de Flavius Philostrate, représentant de la nouvelle sophistique, le dialogue entre Apollonios et Damis sur les nuages "qui courent au hazard à travers le ciel sans rien représenter" et dont nous, "portés comme nous sommes à l'imitation, imaginons et créons ces images".[6]

Ainsi la notion traditionnelle de *mimesis* dans laquelle le peintre ou le sculpteur sont définis comme des artisans parmi d'autres, est remplacée par *phantasia* traduite en latin dans ce cas par *ingenium*.[7]

Plus loin dans ce même dialogue, à une demande formulée à propos d'une influence divine, ou d'une autre qui aurait inspiré Phidias et Praxitèle et qui ne saurait être que l'imitation, il est répondu: "L'imagination, a réalisé ces chefs-d'oeuvres, c'est un artiste dont la sagesse dépasse de loin celle de l'imitation [. . .] l'imagination fabriquera ce qu'elle n'a pas vu pour s'élever jusqu'à l'être et si l'imitation souvent recule effrayée [. . .] l'imagination marche sans crainte vers le but qu'elle s'est elle-même fixé."[8]

Aristote lie la vue à *phantasia* et propose une étymologie de *phantasia* à partir de *phaos*, la lumière car sans lumière il est impossible de voir".[9] Jackie Pigeaud[10] propose le mot français "apparition" avec toutes ses connotations pour traduire *phantasia*. Mais par ailleurs pour Saint Thomas, *phantasia* et *imaginatio* sont synonymes.[11] Cette peinture enfin, gardienne de la mémoire du peintre va être confrontée à un nouvel imaginaire, le regard qui sera porté sur elle et lui donnera son statut "d'oeuvre d'art", regard qui prend lui aussi sa "source" dans ce que dit Aristote. "Nous avons plaisir à contempler les images les plus exactes de choses dont la vie nous afflige dans la réalité" (*Poétique*, 48 b10).

Je propose donc là d'utiliser l'allégorie en montrant comment derrière un sens, la *storia*, se cache la *phantasia* au sens de l'imaginaire; comment dans l'aura qui entoure une oeuvre entrent en écho divers imaginaires: la *storia*, l'imaginaire du peintre, lieu "d'apparition" de son imagination, enfin l'ensemble des regards qui constituent un support d'imaginaires, toujours en mouvement, dans un entre-deux mondes, dès avant ce *hic et nunc* déjà futur.

Je rappelle que l'allégorie est un trope, donc une figure de l'écart par quoi la signification d'un mot est déplacée par rapport à son usage codifié. Définie en terme de mouvement, l'allégorie est la transposition d'un nom qu'Aristote appelle étranger, (*allotrios*) c'est-à-dire "qui désigne une autre chose", 1457 b7. *Poétique* (trad. Hardy, Belles-Lettres) – Barthes dans l'*Aventure Sémiologique*[12] indiquera que dans les mots "de national/étranger et normal/étrange", l'opposition a glissé à propre/figuré.

Dans la *Logique de Port Royal*[13] au mot correspond une signification propre, idée principale et des significations supplémentaires, idées accessoires que Louis Marin dans sa *Critique du Discours*[14] analyse.

Les hommes ne considèrent pas souvent toute la signification des mots.
Les mots signifient souvent plus qu'il ne semble.
On ne représente pas toute l'impression qu'ils font dans l'esprit.
(p. 116 *Logique* . . .).

Louis Marin écrit à ce propos: "Les significations des signes sont donc, primitivement et originairement, des significations propres et toute l'analyse de l'idée accessoire vise à introduire la signification impropre dans un discours théoriquement construit pour ne conjuguer que les premières".[15] Ainsi ces idées accessoires constituent un système décroché de connotations qui peut évoquer, mais en s'en différenciant cependant, le système associatif qui fonde la technique psychanalytique.

Pour Dumarsais, le sens propre est la première signification du mot. Un mot est pris dans le sens propre lorsqu'il signifie ce pourquoi il a été premièrement établi; le sens figuré, un mot pris dans un autre sens qui paraît alors sous une forme empruntée, sous une figure qui n'est pas sa figure naturelle, c'est-à-dire celle qu'il a eu d'abord. Je renvoie ici au *Traité des tropes* de Dumarsais:[16] "La liaison qu'il y a entre les idées accessoires, je veux dire entre les idées qui ont rapport les unes aux autres, est la source et le principe des divers sens figurés que l'on donne au mot."

Cette explication détaillée de l'origine du sens figuré anticipe selon moi, mais évidemment sans qu'il le nomme, le processus de l'association décrit par Freud.

Le terme d'association, emprunté à l'associationisme, désigne "toute liaison entre deux ou plusieurs éléments psychiques, dont la série constitue une chaine associative composée de véritables réseaux correspondant selon Freud[17] à une organisation complexe de la mémoire qu'il compare à un système d'archives ordonné selon différents modes de classement et que l'on pourrait consulter selon différentes voies.

D'où la surdétermination fréquente d'un symtôme pour Freud, mais aussi de toute production psychique, et en particulier verbale ou visuelle ce qui suppose que la représentation (*Vorstellung*) ou la trace mnésique (*Errinerungsspur*) d'un même évenement puisse se retrouver dans plusieurs ensembles que Freud nomme "systèmes mnésiques" en relation entre eux.[18]

Sans le développer ici, je signale l'intérêt de mettre en regard le processus de la théorie des associations étudié par Freud, les idées accessoires dans la *Logique de Port Royal* et le texte de Dumarsais.

Quelle place enfin l'allégorie a-t-elle aujourd'hui dans l'opinion d'un auteur comme Paul de Man? Dans *Allégories de la lecture*,[19] dans l'article "Sémiologie et Rhétorique", Paul de Man souligne que "l'interprétation du signe n'est pas, pour Peirce, une signification mais un autre signe; c'est une lecture et non un décodage, et cette lecture doit être, à son tour, interprétée en une autre signe, ce procédé se répétant ainsi *ad infinitum*. Peirce nomme ce processus, par lequel un signe donne naissance à un autre, la rhétorique pure, par opposition d'une part à la grammaire pure, qui postule la possibilité de signification dyadique et non problématique, et d'autre part à la logique pure, qui postule la possibilité de la vérité universelle des significations".

Au-delà de la figure de rhétorique qu'est l'allégorie, au-delà de sa valeur instrumentale, elle est donc réactualisée aujourd'hui par les sémiologues qui en font une dynamique pour produire *ad infinitum*. Ce *ad infinitum* marque ici une propriété – l'infini du point de fuite en géométrie – et non pas une finalité. Il permet d'arrêter, comme en peinture, la limite du cadre que l'on détermine. Ainsi ce volume d'infini serait cadré, à chacun des arrêts que l'on choisit d'opérer, définissant alors, chaque fois, un champ sémantique différent.

Le tableau de Birmingham fait fonctionner un paysage héroïque. Par paysage héroïque, j'entends le paysage dans son sens propre c'est-à-dire la description de convention du paysage épique tel que le décrit Curtius:[20] la mise en place topographique du paysage, par exemple: le *positus locorum* dans Stace (*Silves* V.3.236), disposition des lieux; le *situs terrarum* dans Horace (*Epitres*, II, 1.25), position de la région.

L'exigence de précisions topographiques parcourt le poème héroïque dont: la forêt sauvage, *una selva selvaggia ed aspree a forte* dans Dante, *la forêt aux essences mélangées* chez Homère et Virgile; le *locus amoenus* (qui peut même se trouver dans une forêt sauvage: Curtius mentionne la vallée de Tempé en Grèce) sont particulièrement caractéristiques. J'entends aussi par paysage héroïque son sens figuré, au sens où les différents éléments d'un tableau peuvent composer une figure de l'héroïque.

Le Tableau de Birmingham a été peint vers 1636–37 et fait suite d'environ cinq-six ans au tableau actuellement à Saint Pétersbourg qui montre le même passage de *la Jérusalem Délivrée* de Torquato Tasso, publiée en 1581:[21] Herminie accompagnée de Vafrin découvre aux abords de Jérusalem, le corps inanimé de Tancrède. Le tableau de l'Ermitage représente la peinture d'un passage d'une *storia.* Ici nous sommes dans le droit fil du texte du Tasse, strophes 111–112, Vafrin voyant que Tancrède n'est pas mort lui enlève son armure et Herminie coupe ses cheveux pour étancher ses plaies.

Dans le tableau de Birmingham, le paysage a changé: il est empreint d'un merveilleux qui ne règnait pas ainsi dans la première version, on pourrait presque dire que ce tableau se meut dans un ailleurs où s'allégorise, où Poussin allégorise la *storia* qu'il avait représentée dans le tableau de l'Ermitage. Tout se passe comme si le tableau représentait une scène théâtrale.

Dans un sous-bois accidenté, nous sommes dans le désert de la forêt, dans ce désert-forêt si bien décrit par Jacques Le Goff, dans l'*Imaginaire médiéval*,[22] huit personnages sont les protagonistes de l'action. Tancrède, au premier plan, est soutenu par Vafrin qui n'est plus agenouillé comme dans le tableau de l'Ermitage, mais en équilibre dans le mouvement.

Herminie agenouillée cette fois devant lui est en train de trancher sa chevelure d'un geste décidé. Vafrin debout, Herminie agenouillée sont peints dans des attitudes inverses de celles du premier tableau: Vafrin agenouillé, Herminie debout. Au-dessus de ce groupe de trois personnages, dans la partie gauche du haut du tableau, deux *putti* volent, symbolisant du moins. Voudrais-je le proposer, l'amour d'Herminie pour Tancrède, peinture de sa vertu – et peut-être tels un écho de cette *candor* blancheur étincelante – qui illumine et Herminie et le cheval blanc dans le tableau de l'Ermitage.

De manière plus diffuse, cette lumière éclaire par endroit Vafrin, et fait étinceler sa cuirasse et son casque. Enfin elle touche la crinière et la queue des deux chevaux, leurs jarrets, la gorge, les épaules, le dos, la tête qui regarde ailleurs du cheval de droite et elle accroche les reflets des armes qui occupent le coin droit inférieur du tableau. Cette lumière

Nicolas Poussin, 'Tancred and Erminia'. Collection of The Barber Institute of Fine Arts, The University of Birmingham.

semble émaner des deux *putti*, eux-même baignés de clarté, qui regardent avec une ineffable tendresse la scène qui se joue au-dessous d'eux. Si j'insiste sur cette lumière, c'est que les deux *putti* dont je vais montrer la fonction dans ce paysage héroïque, en sont la source.

Les lointains se fondent dans le ciel crépusculaire: le soleil s'éteint à l'horizon et au loin, à peine distincte, sur la droite du tableau, derrière le cheval qui détourne la tête, une vague indication de Jérusalem s'esquisse.

L'espace du premier plan – différemment du terrain vide du tableau de Saint Pétersbourg – est totalement saturé: là où fleurissaient quelques plantes près d'une anfractuosité de rocher, le casque de Tancrède, sa cuirasse, son sabre et son bouclier sont rassemblés, dans un effet de "nature morte" qui occupe tout l'angle inférieur droit du tableau.[23] Poussin a admirablement mis en valeur les armes, emblèmes de ce qui est l'essence de l'épopée et qui parcourt tout le poème du Tasse: la guerre.

"Mais la grandeur de l'épopée est fondée sur les exploits d'une très haute vertu guerrière".[24]

La Jérusalem est le récit d'un combat au sens propre, au sens littéral; d'un combat amoureux au sens figuré, allégorique, voire d'une sublimation religieuse.

L'atmosphère du tableau de Birmingham suggère l'enchantement de la forêt d'Armide, sa magie. Cette scène issue de la réalité – un homme inanimé est secouru par un homme et une femme – se situe, par le pinceau de Poussin, dans une autre dimension qui est commensurable avec l'oeuvre du Tasse: l'allégorie.

Dans le tableau de l'Ermitage, Poussin semble jouer sur le contraste entre un paysage statique, mis en place topographiquement selon le texte du Tasse, des personnages immobiles et Herminie, personnage central de ce passage du poème, autour de qui se concentre le mouvement. Dans le tableau de Birmingham, au contraire, tout n'est que mouvement. Les chevaux impatients attendent la fin de l'épisode, déjà tournés vers le suivant, tournés vers Jérusalem. Le paysage lui-même s'anime comme une forê vivante de rochers mouvants, est-on tenté de dire: c'est la forêt magique d'Armide.

C'est dans la version de Birmingham que le travail sur les étoffes et les plis des draperies est le plus remarquable, donnant son plein pouvoir émotionnel à la science du coloris. La tunique d'Herminie, l'étoffe jetée sur Tancrède se répondent l'une à l'autre dans leurs tonalités fortes, le rouge et le bleu.

Cette retenue dans l'art de disposer ces plis, de les colorer et, dans la couleur de les revêtir d'ombre et de lumière nous semble suivre Franciscus Junius "*color et in eo lux et umbra*"[25] et répondre à Lomazzo qui place si haut le pouvoir de l'émotion créé par les qualités du coloris.

Les deux chevaux regardent ailleurs, vers le lointain de droite, hors champs en piaffant. La forêt où nos nous trouvons est bien cette forêt profonde que nous peint le Tasse et, où la lumière, telle celle d'un ciel brouillé n'éclaire que très peu. Une lumière, dont la source n'est pas "naturelle" se projette sur le torse de Tancrède – sa tête, ses épaules sont dans l'ombre – le bas de son corps, l'étoffe dont il est drappé et les parties de son corps qui touchent ceux qui le secourent: sa main gauche qui s'appuie sur le genou droit d'Herminie, placée dans une position centrale dans le tableau, son avant-bras et sa main droits qui s'appuient sur la cuisse droite et la cuirasse de Vafrin. Cette même lumière éclaire le visage, le haut du corps, le pied gauche et par endroit le vêtement d'Herminie.

Cette nouvelle Herminie inventée par Poussin est dans ce tableau une héroïne dont la chair est exposée, on pourrait dire idéalisée par la lumière qui l'éclaire et, renonçante: les *putti* symbolisant son renoncement. Je vais y revenir. Tancrède apparait dans une posture qui renvoie à la *Déposition de l'Ermitage* (autour de 1620): dénudé mais chaste.

Ainsi le paysage de Birmingham apparaît comme un paysage allégorisé dans la multivalence de l'image, paysage intérieur des sentiments traversant le poème qui semble nourrir le tableau.

J'aborderai quatre motifs:

La couleur de la lumière

Dans le tableau de l'Ermitage, les lointains laissent la plus grande partie au ciel "l'orient brunit au coucher du soleil". C'est à la fin du jour que Vafrin et Herminie trouveront Tancrède inanimé.

Dans ces deux tableaux la lumière a une fonction essentielle. Dans le tableau de l'Ermitage, elle désigne et condense, dans le tableau de Birmingham elle emblématise. C'est une lumiére surnaturelle qui dans le tableau de l'Ermitage éclaire Herminie et le cheval blanc. Il s'agit vraiment de "candor", d'une lumière d'une blancheur éclatante, qui dans les mots du Tasse au chant VI, entoure Herminie qui s'est parée de l'armure de Clorinde: "elle était en un point où les beaux rayons célestes frappent d'aplomb son armure si brillante qu'on aperçoit de loin l'éblouissante blancheur qui l'entoure et l'enveloppe" (VI-106).

Poussin qui s'inspire du chant XIX du poème du Tasse pour son tableau rend traversante cette *candor* qui éclaire Herminie dès le chant VI. Poussin, on le sait par sa lettre à Chambray du 1er mars 1665,[26] commentera le traité de Junius qui paraît à l'époque où il peint les deux versions de *Tancrède et Herminie*. Dans cette lettre à Chambray, il écrit: "c'est une imitation faite avec lignes et couleurs en quelque superficie de tout ce qui se voit dessous le soleil, sa fin est la délectation". Et quel tableau lumineusement solaire que celui de Saint Pétersbourg, différent de celui de Birmingham où nous sommes dans l'enchantement allégorique d'un sous-bois.

Les putti comme emblème

La peinture des *putti* révèle la recherche d'une *maniera greca*, d'une douceur en peinture ainsi que l'exprime Charles Dempsey dans son essai sur *The Greek Style and the Prehistory of Neoclassicism.*[27] Il fait remarquer l'importance qu'eut pour Duquesnoy, pour Boselli, pour Pietro Testa et Poussin cette manière grecque que leur révélèrent la statuaire antique et l'étude toute particulière de la collection Ludovisi: le but d'une telle étude était d'acquérir une grande manière, définie non par son côté grandiose, sa noblesse, sa majesté mais par une gracieuse légèreté (*leggiadria*), une douceur et une tendresse exprimées dans la délicatesse d'une gradation imperceptible dans le contour.

Passeri, dans sa vie de Duquesnoy insiste sur la quête de Poussin et Duquesnoy de cette "*maniera greca*, la manière maitresse de l'art parce qu'elle contient en elle, en même temps, la grandeur, la noblesse, la majesté et la grâce qu'il est difficile de réunir en une seule oeuvre".[28]

Poussin et Duquesnoy s'inspirent des *putti* peints par Titien dans son *Adoration de Venus* en 1519; Poussin dessinera une *Fête de Venus* et ses célèbres *Baccanales de Putti* (Palazzo Barberini à Rome), qu'Antony Colantuono,[29] nomme à la suite de Bellori "scherzi e baccanali" en écho aux poèmes de Marino dont Poussin avait illustré l'Adone. Duquesnoy sculptera autour de 1626–27 ses bas-reliefs, le *Triomphe de l'amour divin sur l'amour profane* et la *Baccanale avec une chèvre.*

Poussin et Duquesnoy découvrirent un nouveau *putto* dit Orfeo Boselli.[30] Le *putto moderno* est plus jeune que le *putto antico*. Ce *putto antico* employé par les anciens était montré à l'âge convenant à l'accomplissement de ce qu'il était en train de faire.

Boselli loue les anciens pour leur sens de la *convenienza* (convenance) et fait remarquer que Raphaël suivait cette ancienne manière de peindre les *putti*. Comme les anciens il les montrait suffisamment âgés pour accomplir les actes qu'on leur demandait d'accomplir. On peut observer la même *convenienza* dans les *putti* qui conduisent les chars triomphaux de Frédérigo Montefeltro et Battista Sforza, peints par Piero della Francesca après 1472 (Florence, Office); chacun des *putti* est capable physiquement de guider le char.

La figure du *putto moderno* s'identifie au concept métaphorique d'une qualité de douceur c'est-à-dire que le *putto* est une visible et particulière manifestation de l'idée de douceur. Ses proportions, l'essence de sa capacité à incarner l'idée de douceur sont aimables.

Dans le tableau de Birmingham la figure des *putti* donne une forme visible au *concetto*. L'usage du terme *concetto* nous renvoie à un autre terme avec lequel il entretient des rapports d'affinité: l'*impresa* "cette image-idée, noeud de paroles et d'images [. . .] était en effet, aux yeux de plusieurs, le *concetto* à l'état pur".[31]

Dans ces tableaux, celui de Birmingham en particulier, on ne peut s'empêcher de tisser des liens secrets entre les personnages: les *putti* liés à Herminie semblent là peints pour donner corps aux *affeti* et quels corps délicats et subtils "dans un suave accord de lumières et d'ombres" ainsi que le souhaite Poussin dans son propos sur l'idée de beauté.[32]

Rappelons très schématiquement le parcours de l'idée de beauté tel qu'il est tracé par Panofsky dans *Idea*,[33] où s'exprime la position néo-platonicienne de Poussin vis-à-vis de ce concept.

Bellori[34] cite Poussin qui s'inspire de Lomazzo dans l'*Idea del Tempio della Pittura*,[35] "De la façon de connaître et de construire les proportions suivant la beauté" (chap. 26), qui, lui-même s'était inspiré de Marsile Ficin dans le *Commentaire au Banquet de Platon* – (Florence 1544 au début du chapitre VI) "Combien de parties concourent pour faire une chose belle et que la beauté est un don spirituel". Poussin fait siens les propos de Lomazzo sur l'astrologie et ceux de Marsile Ficin sur la musique.

Le sens du tableau joue ainsi entre plusieurs éléments – figures, paysages, couleurs – disposés dans une différence étudiée par rapport à la *storia*, et dont la découverte du sens est laissée à notre intention, à notre plaisir. Et l'usage de la figure du tendre *putto* dans le tableau de Birmingham incarne bien le *concetto*, "l'image-pensée" de l'amour d'Herminie. Si le tableau de l'Ermitage est une peinture d'histoire suivant la *storia*, celui de Birmingham qui se nourrit de moments très différents du poème, se décale dans un intervalle qui est en rupture avec le texte du Tasse dans son déroulement chronologique, fracture de la cohérence du système de représentation, distorsion qui l'investit alors d'une nouvelle signification où nous lisons au-delà de l'allégorie du poème, l'intention allégorique de Poussin.

L'amor hereos

Ces tableaux condensent le personnage d'Herminie tel qu'il apparaît tout le long du poème: elle est bien cet *amor hereos*[36] qui aime d'un amour héroïque et divin. Giordano Bruno dans les *Fureurs Héroïques* écrit: "Son amour, cette fois, est pleinement héroïque et divin; et je le veux entendre ainsi, encore que par cet amour dit-il, lui soient infligées si cruelles tortures; car tout amant, désuni et séparé d'avec l'objet qu'il aime (auquel, conjoint en affection, il voudrait l'être en effet) se trouve en deuil et peine, se tourmente, se met au supplice, non certes parce qu'il aime, ayant conscience que son amour est très dignement et très noblement appliqué, mais parce qu'il est privé de cette fruition qu'il obtiendrait s'il était parvenu au terme auquel il tend. Il ne souffre pas du désir qui le fait vivre, mais des difficultés qu'oppose à son ardeur, une épreuve qui lui fait souffrir le martyre".[37]

Herminie présente tous les symtômes de cette maladie: plongée dans des rêveries secrètes elle pousse de douloureux soupirs, souhaitant qu'un jour, dans cette forêt, vienne son héros (Chant VII-21). Giordano Bruno exprime de façon exemplaire cet *amor hereos* qui anime Herminie et la conduit à la sublimation de son amour: "Car ce mal, dans l'absolu et au regard de l'éternité, est tenu soit pour bien, soit pour guide conduisant au bien, attendu que ce feu est l'ardent désir des choses divines".[38]

Sur l'art de peindre avec les mots

Orazio Lombardelli dans une lettre à Maurizio Cataneo du 28 Septembre 1581, publiée par le Tasse, dans l'*Apologie en défense de la Jérusalem délivrée*, en 1585,[39] fait l'éloge du Tasse comme poète toscan – il a écrit ses vers en italien – et comme défenseur de l'Eglise: l'exaltation de la foi parcourt tout le poème. On pourrait dire que le Tasse peint en poésie, dans ses descriptions, les lieux, les temps différents: la nuit et le jour, les orages et les accalmies, le réel et l'irréel, les batailles, les animaux, les essences des arbres, ornant ses "tableaux" de florilèges pour enchanter l'esprit.

A la fin du IIe chant, lorsqu'Argan décide d'entrer dans Jérusalem, la description de la nuit et du repos général des éléments, des animaux, va permettre de souligner l'état de veille des croisés, veille qui trouvera son écho dans la peinture de l'Aurore.

Le Tasse, dans son poème, va et vient d'un réel – le monde "naturel" – à un irréel, le monde magique: "Le ciel redevint limpide, l'air paisible la forêt revint à son état naturel: non point délicieux ou terrible par magie, mais pleine d'une horreur qui est son horreur innée" (chants 18-38).

Ainsi la forêt, le paysage s'enchantent devant nos yeux, le poète peint pour nous cet ensorcellement: aux chants 18-22-23. "Mais le désir de nouveauté qui le pousse (il s'agit de Renaud) à explorer l'antique et profonde forêt, et de merveille en merveille l'entraîne dans ses solitudes sauvages. Où qu'il aille on dirait que tout jaillit, tout fleurit sous ses pas: là s'ouvre un lys, ici pousse une rose, ici naît une source, là coule un ruisseau et tout alentour et au-dessus de lui la forêt séculaire semble rajeunir toutes ses feuilles."

La forêt séculaire, l'antique forêt est un autre exemple du paysage idéal, formé par Virgile dans l'Enéide. (Je renvoie ici à Curtius.)

Le choix et la connaissance d'une belle matière énoncés dans ce propos se rapprochent de ce qu'écrit le Tasse dans *Le premier discours sur l'art poétique.* Il y fait une comparaison entre la matière nue qui n'a pas encore été . . . travaillée par le poète, et la nécessité pour un constructeur de bateaux de savoir, au-delà de la forme qu'il compte donner à son bateau, de quelle manière le bois dont il sera construit est le plus apte à recevoir telle ou telle forme". Conception aristotélicienne de l'art poétique.

Dans le tableau de l'Ermitage, ce que Poussin nous fait voir s'apparente pourrait-on dire au contenu manifeste: l'histoire qui nous est montrée ici, en peinture, n'est pas à double entente. Le "je ne sais quelle lumière"[40] dans ce tableau, c'est la lumière du soleil couchant. Je voudrais proposer une version divine et biblique du poème. Alors que ce que peint Poussin dans le tableau de Birmingham s'apparente pourrait-on dire, au contenu latent, tel le "un jeune sais quoi de lumineux" emprunté au chant XVII-57 du Tasse. En effet, la lumière d'un soleil couchant, au loin, à l'horizon, n'a aucun rapport avec la lumière "inexplicable" qui éclaire les *putti*, lumière qui a une logique interne: le *putto* de gauche jette une ombre sur le *putto* de droite dont il masque une partie du corps. Les *putti* semblent être eux-même source de lumière, lumière qui venue d'en haut à gauche rencontre des obstacles. Dans le tableau de l'Ermitage, la source de lumière serait hors-champ, venant de la gauche et éclairant en particulier, de manière enveloppante, le cheval blanc, le témoin de la scène.

Dans l'analyse d'un rêve, le contenu manifeste est le rêve tel que le rêveur en fait le récit. Par extension on peut parler du contenu manifeste de toute production. Le contenu latent s'exprime, lorsqu'on fait l'analyse du rêve, dans l'ensemble des significations auxquelles cette analyse aboutit. Le contenu manifeste s'exprime tel un récit en images qui, une fois déchiffré, permet l'accès au contenu latent où l'on atteindra une organisation des pensées, un "discours". Dans l'*Interprétation des rêves*, Freud écrit: "Les pensées du rêve et le contenu du rêve nous apparaissent comme deux exposés des mêmes faits en deux langues différentes; ou mieux, le contenu du rêve nous apparaît comme une transcription (*Übertragung*) des pensées du rêve, dans un autre mode d'expression".[41] "La parole du poète, telle qu'elle se développe dans l'activité poétique, est solidaire de deux notions complémentaires: la Muse et la Mémoire. Ces deux puissances religieuses dessinent la configuration générale qui donne à l'*Alétheia* poétique sa signification réelle et profonde".[42]

L'*Alétheia*: La Vérité, est une notion ambigüe qui, "double",[43] est connotée territorialement par l'intermédiaire de la *mimesis* et qui se situe entre deux pôles: *Alétheia* qui est le côté positif, *Léthé* le côté négatif.[44] Entre *Aléthéia* et *Léthé*, vérité/oubli, dans l'intervalle de leur distance que pourrait être la mémoire – un imaginaire, ce que l'on fait d'un réel – se construit un irréel que l'on pourrait placer en face de ce réel. La construction "in absentia" de ce monde intérieur par le peintre trouverait ainsi son réel, sa vérité. Le poète, dans la Grèce archaïque, tel un visionnaire, accède directement "par sa mémoire [. . .] dans une vision personnelle, aux événements qu'il évoque; il a le privilège d'entrer en contact avec l'autre monde. Sa mémoire lui permet de déchiffrer l'invisible".[45]

Dans le tableau de Birmingham, dans sa représentation d'un épisode de *La Jérusalem*, au-delà de l'allégorie, Poussin nourrit sa peinture d'un imaginaire qui lui est propre. On pourrait proposer comme "clef" de ces deux tableaux, l'éternel souci, l'éternelle mémoire que Poussin porte en lui des ombres et de la lumière utilisées si différemment, comme nous venons de le voir, dans le tableau de l'Ermitage et dans celui de Birmingham où la magie règne et où Poussin montre, dans son talent de peintre, par l'intermédiaire de ce qui en fait matériellement l'emblème, ici – les *putti*[46] – sa recherche de la douceur en peinture.

Ecole d'Architecture de Versailles

NOTES

[1] E. Cassirer, *Individu et cosmos dans la philosophie de la Renaissance* (Paris: Minuit, 1983, Réimp. de Leipzig et Berlin 1927), pp. 172–173.

[2] G. Couton, *Ecritures codées, Essais sur l'allégorie au XVIIe siècle* (Paris: Aux Amateurs de livres, 1990).

[3] E. Bloch, *Experimentum mundi* (Paris: Payot, 1981), pp. 194–195.

[4] Quintilien, *De Institutione Oratoria*, Livre VIII. ch. 6, 44 (Paris: Belles Lettres), p. 116.
[5] J. Pepin, *Mythe et Allégorie* (Paris: Aubier, 1958), p. 86.
[6] Reinach (1.1.) n° 52, *Recueil Millet*, Macula.
[7] B. Schweitzer, "Mimesis und Phantasia", *Philologus*, **89**, 1934, p. 286 sq.
[8] *Ibid.*
[9] *De anima*, 429 a.
[10] *Folie et cures de la folie chez le médecins de l'antiquité gréco-romaine, la manie* (Paris les Belles Lettres, 1987), p. 95.
[11] R. Klein, *La forme et l'intelligible* (Paris: Gallimard, 1970), p. 65.
[12] R. Barthes, *L'Aventure Sémiologique* (Paris: Seuil, 1985).
[13] *La Logique ou l'Art de penser contenant outre les règles communes plusieurs observations nouvelles propres à former le jugement*, 5e édition (Paris: Desprez, 1683).
[14] L. Marin, *Critique du Discours* (Paris: Minuit, 1975).
[15] L. Marin, *Critique . . . , op. cit.*, p. 314.
[16] Dumarsais, *Traité des Tropes*, 1730 (Paris: La Connaissance, 1977) p. 28.
[17] Je renvoie aux *Etudes sur l'Hystérie* de Freud et Breuer, trad. Berman, 1985 (Paris: PUF, 1956).
[18] Je renvoie à quelques travaux princeps de Freud dans ce domaine. *Etudes sur l'Hystérie*, 1895 (Paris: PUF, 1956); *Lettre à Wilhelm Fliess* du 6.12.86 in *Naissance de la Psychanalyse* (Paris: PUF, 1956); Le chapitre VII de l'*Interprétation des rêves*, 1900 (Paris: PUF, 1973) pp. 456–461.
[19] P. de Man, *Allégories de la lecture*, 1979 (Paris: Galilée, 1989).
[20] E. R. Curtius, *La Littérature européenne et le Moyen Age latin*, 1 (Paris: PUF) pp. 323–326.
[21] Le Tasse, *La Jérusalem délivrée*, 1581 (Paris: Garnier Flammarion, 1990).
[22] J. Le Goff, *L'imaginaire médiéval* (Paris: Gallimard, 1985) p. 64.
[23] Pour qualifier les objet – un livre, un luth, une partition de musique et une palette de peintre – qui se trouvent au pied de Mercure dans le tableau de Poussin: *Venus et Mercure* à la Dulwich Picture Gallery, R. Verdi emploie la même expression "still life", in N. Poussin, *Venus and Mercury* (Dulwich Picture Gallery, 1986).
[24] Tasso, *Scritti sull'arte poetica*, Discorso primo (Einaudi Turin, 1977), p. 14.
[25] Franciscus Junius, *De pictura Veterum* (Amsterdam: Iohanes Blaeu, 1637, Livre III), p. 137.
[26] N. Poussin, *Lettres et propos sur l'art*, textes réunis par A. Blunt (Paris: Hermann, 1989), pp. 173–174.
[27] Ch. Dempsey, "The Greek Style and the Prehistory of Neoclassicism", in E. Cropper, *Pietro Testa, 1612–1650, Prints and drawings, with essays of Ch. Dempsey* (F. Solinas and Anna Nicolo, Francesca Consagra: Philadelphia Museum of Art, 1988).
[28] G. B. Passeri, *Vite de'pittori, scultori ed architetti che hanno lavorato in Roma, de 1641 al 1673* (Rome, 1772), pp. 110–112.
[29] A. Colantuono, *The Tender Infant: Invenzione and Figura in the Art of Poussin* (Thèse: Hopkins, 1986).
[30] *Osservazioni della scoltura antica dai Manoscritti Corsini e Doria e altri scritti*, éd. Phoebe Dent Weil (Florence: S.P.E.S., 1978), IV, 16, Fol. 124 v.
[31] R. Klein, *La forme et l'intelligible* (Paris: Gallimard, 1970), p. 130.
[32] N. Poussin, *Lettres et propos . . . , op. cit.*, sur l'idée de beauté, p. 183.
[33] E. Panofsky, *Idea* (Paris: Tel, Gallimard, 1989), pp. 254–255.

[34] G. P. Bellori, *L'idée du peintre, du sculpteur et de l'architecte, tirée des beautés naturelles et supérieures de la nature* in *Idea . . . , op. cit.*
[35] G. P. Lomazzo, *Idea del tempio della pittura* (Milano, 1590).
[36] Sur la sémantique de *amor hereos*, voir G. Agamben, *Stanze, Parole et fantasme dans la culture occidentale*, traduit de l'italien par Y. Hersant (Paris: Bourgois, 1981), IIIe partie, chapitre V, pp. 185–195.
[37] G. Bruno, *Des fureurs héroïques*, 1585, traduction de P. H. Michel (Paris: Les Belles Lettres), Première partie, Dialogue 1, pp. 186–188.
[38] *Ibid.*
[39] B. Weinberg, *A History of Literary Criticism in the Italian Renaissance*, Vol. II (Chicago: The University of Chicago Press, 1961), pp. 983–990, pp. 1009–1059.
[40] Emprunté au chant XVII, 57: " . . . non so che di luminoso . . . ", Tasso, *Gerusalemme liberata* (Milano: Garzanti), p. 517.
[41] S. Freud, *L'interprétation des rêves*, tr. I. Meyerson (Paris: PUF, 1926 et 1967) p. 241.
[42] M. Detienne, *Les maîtres de vérité dans la Grèce archaïque* (Paris: Maspéro, 1981), pp. 10.
[43] *Ibid.*, p. 26.
[44] *Ibid.*, pp. 76–77.
[45] *Ibid.*, p. 15.
[46] Nous faisons ici, les *putti*, héritiers de l'Eros, petit enfant de la fin du IVe siècle avant J.C. à l'âge hellénistique.

Ce texte est extrait pour partie de la thèse menée a l'École des Hautes Études en Sciences Sociales sous la direction d'Yves Hersant et intitulée: Le Sens voilé, défenses et illustrations de l'allégorie.

LOIS OPPENHEIM

MICHEL BUTOR'S *L'EMBARQUEMENT DE LA REINE DE SABA*: AN ALLEGORICAL UNDOING OF AESTHETIC EXPERIENCE

Modern critical studies and the meta-critical approaches to art and literature to which they adhere are increasingly divisible into two philosophic camps: the foundationalist and the contextualist. "Each may take a variety of forms," notes Gary Shapiro in his essay "Gadamer, Habermas and the Death of Art,"

> but it is typical of the foundationalist to seek a fixed basis for . . . the claims of aesthetic and artistic experience, while the contextualist denies the availability of any such fixed grounds and suggests instead that . . . our aesthetic experience can get along well enough with reference to the norms and practices of actual societies, taking into account the historical experience of these societies as it is found in their cultures.[1]

Not only in his critical works (of which there exist the five volumes of essays known as *Répertoire* and several miscellaneous volumes on literature and painting), but in his creative works as well, Michel Butor aims – through textual correlations, filiations, interconnections – to reveal paradigms that, in yielding clues to the history and actuality of a given culture, order, in some sort of comprehensible totality, the formlessness and ambiguities of life. His efforts to render the world and the diversity of its cultures intelligible derive from a vision that is fundamentally relational – not contrastive, but connective, one that relativizes what many have taken, over the years, to be purely formalist concerns: His play with numbers and with quotations (from his own texts as well as those of others) and his generally remarkable manipulation (classification, repetition and so on) of all kinds of material seemingly extraneous to the advancing of the narrative line serve to contextualize the otherwise constructivist character of his writing.

Although he neither theoretically nor creatively adheres to the notion of so-called intrinsic qualities in a work (his own or another's) that would qualify it as "art" (in the institutionalized sense) however, Butor does demonstrate in his writing a non-relativistic functioning of art that manifests itself in self-reflexivity. This is to say that, though he consistently disassociates art and literature from the purity of aesthetic experience defined by foundationalism, his focus on the commplace,

M. Kronegger and A-T. Tymieniecka (eds.), Analecta Husserliana XLII, pp. 131–139.

his privileging of the banal and all that we generally take for granted in our daily lives, not only effects a *tabula rasa* of aestheticism (in a manner not unlike that of a Duchamp or a Warhol), but reflects the composition of the narrative (and hence the writer at work) and the production of meaning (and hence the reader at work) as well. Nevertheless, in a paradoxical reversal characteristic of all Butor's writing, this self-reflexivity exemplifies nothing other than the author's belief in the cultural situatedness of every artistic enterprise and reveals, once again, that any notion of a pure aesthetic experience is fundamentally anathema to him. In what follows, I will argue that it is precisely the allegorical function of this writing – evident not only in the positing of the banality of the object, but also in the text's self-reflexivity – that grounds Butor's work in contextualism and calls for our situating it in an anaesthetic (as opposed to properly aesthetic) arena.

Allegory is viewed by some as a powerful medieval and Rensaissance form of little significance for the reader of modern or contemporary texts and by others as "omnipresent in Western literature from the earliest times to the present."[2] Such variation in the appraisal of allegory and its relevance to the understanding and appreciation of recent writings would suggest that there is little agreement as to those features that define its perimeters and delimit its function. Indeed, one would agree with Angus Fletcher who, in his comprehensive analysis of allegory, calls it "a protean device"[3] although its variability should not eliminate the possibility of consensus as to its most essential characteristics. A general survey of attempts to define allegory, however, reveals that it is the profound dichotomy inherent in contemporary critical approaches to art and literature that impedes unanimity.

Roland Barthes has written of allegory as a meaning that brings together a signifier and a signified,[4] that which conforms to his view of language as divisible from that of which it speaks. Similarly, Paul de Man, in a highly questionable attack on Serge Doubrovsky's *Pourquoi la Nouvelle Critique?*, denies perception as a possible model for literary creation and links allegorical with representational poetics.[5] Subscribing to the notion that allegory, in saying one thing while meaning another, is "abnormal language,"[6] numerous pre-poststructuralist critics, like their poststructuralist successors for whom "allegory's indirect signification is actually the property of *all language*, for no statement is open and direct . . . ,"[7] to quote Carolynn Van Dyke, distinguish themselves from philosophers and critics who deny the representational premise. Recent poststructuralist criticism, as Van Dyke points out, attempts "to normalize allegory as the condition of all textuality"[8] insofar as it "bases itself frankly on the disruption of signifier and signified and . . . renounces the illusions of semantic unity and directness promoted by such modes as symbolism."[9]

Hans-Georg Gadamer, on the other hand, while recognizing that in allegory, as in the symbol, "one thing stands for another,"[10] avoids contrasting a visually imageable thing and its verbal imaging or representation. Rather, what in allegory stands in a relation of alterity to something else is "the meaningful relation of the sensible to the non-sensible,"[11] the "presentation in images for something that is imageless."[12] Hans Robert Jauss goes so far as to qualify allegorical style as "beauté inutile,"[13] what de Man describes as "the absense of any reference to an exterior reality of which it would be the sign."[14] And Jacques Garelli, like Merleau-Ponty before him, abandons representational thinking in his alliance of signifying structures of all kinds with the sensorial dimension of literary or pictorial works of art.

The question of reference, therefore, the relativity of meaning to and its contingency on the material world, is responsible for the divergent points of view on allegory and on its presence and function in modern literature. If we have chosen the writing of Michel Butor, and *L'Embarquement de la Reine de Saba* in particular, as the focus of this study, it is that the referentiality of art is explored in this work, with the representational value of art significantly devalued, and this within the framework of the allegorization of art itself.

In 1989 Butor published *L'Embarquement de la Reine de Saba*, a work which may best be described as a written corollary of Claude Lorrain's 1648 painting of the same name, a painting that hangs in the National Gallery of London and was inspired by the tenth chapter of the Bible's First Book of the Kings, which begins:

And when the queen of Sheba heard of the fame of Solomon concerning the name of the LORD, she came to prove him with hard questions.

And she came to Jerusalem with a very great train, with camels that bore spices, and very much gold, and precious stone. . . .[15]

Butor's narrative aims not to interpret the Lorrain painting, but rather to *animate* it and to exemplify thereby the dependence of the production of meaning, whether verbal or visual, on reader or viewer experience of the work in question. His effort is to define the relation of reader and text in terms of that between himself and/or his reader as viewer and the Lorrain work and this within the context of perceptualism. It is to the extent that the constitution of the verbal meaning (that of Butor's text) both reflects and completes the constitution of non-verbal meaning (that of Lorrain's painting) on the one hand, and, on the other, to the extent that, in this demonstration of the production of meaning, Butor valorizes the particular as an illustration of the general, that this narrative may be interpreted as an allegorization of art. It appears to function allegorically in a number of other ways as well, however, and these are of interest to us here not for the kinds of meaning that they may or may not reveal (historical, prophetic, moral and so on), but for their utter lack of reliance on the fixity of aesthetic experience. This is to say that the constitution of *textual* meaning *through* the constitution of the *visual* signification of the Lorrain painting – a process that posits allegorically the question of reference – demonstrates the contextualizing as opposed to idealizing or abstracting function of the allegorical structure, and that this contextualization occurs despite the text's reflecting back on itself rather than on any reality exterior to it.

In *L'Embarquement de la Reine de Saba*, as in Butor's well-known 1957 novel *La Modification*, where the protagonist was "vous" (or you) the reader, the contemporaneity of the production of meaning with the perceptual act of reading is allegorized in the journey motif. In the earlier work, the journey objectified, thematically, the movement from pre-cognitive to cognitive awareness, from non-thetic to thetic consciousness, that resulted in the character's decision regarding his relationships with wife and mistress. And the journey objectified, allegorically, the unfolding of textual meaning, the way in which the semantic field unfolds from the reciprocal (or intersubjective) interaction of character, author and reader for the transition from one place to another (from Paris to Rome) told the story of a decision to write the novel we, in fact, were in the process of reading.

While the queen's journey to Jerusalem in *L'Embarquement de la Reine de Saba* is not structured on the same kind of fictive *mise en abyme* as *La Modification*, it does specularize the temporalizing process that generates its meaning – and hence that of the Lorrain work as well – and therein lies the work's allegorization of art. Numerous projections into the figure – in the form of musings by the text's "characters" (inhabitants of course of the painting also) on what King Solomon's palace will be like – contrast with the opening line of each small section of the book which maintains a focus on the "reality" of the depicted scene to illustrate the conditioning of perception by association, memory and the imagination. We read, for example, the following answers to the question "A quoi ressemblera ce palais du roi Salomon?"[16]

> Jacinthe:
> "J'imagine une mer de bronze ornée de deux rangées de coloquintes et reposant sur douze boeufs."
>
> Narcisse:
> "J'imagine que la chemin, pavé de cailloux bleus, craque sous les pas, et que des roses épanouies pendent en berceau sur toute la longueur de l'allée."
>
> Jonquille:
> "J'imagine que les cent portes donnent entrée dans des jardins ou des magasins remplis de richesses, ou enfin dans des lieux qui rêferment des choses surprenantes.[17]

These imaginings suspend in fiction the "realistic":

> La Reine lève le pied gauche sans le découvrir.[18]
>
> La Reine pose le pied gauche sur la marche suivante san le découvrir.[19]
>
> La Reine lève le pied droit au-dessus de l'eau.[20]

It is this dialectical and thus temporal tension on the level of the fiction between the imaginary and the real, a tension that mirrors the verbal surpassing and conservation of Lorrain's pictorial real, that allegorizes art as a simultaneous retention and deformation, a surrealization, in sum, of the material world. Like any creative work, moreover, Butor's text reveals something other than that which is presented in the work and, to the extent that the *verbal* text unfolds in a fictive journey the development of meaning in the *visual* work of art, it allegorizes both its own creative function and the creative functioning of all art. Irreducible to the order of objectivity, the work of art embodies the transcendental dimension of the creative imagination that gave rise to it.

If we prefer to characterize *L'Embarquement de la Reine de Saba* as an anaesthetic rather than a uniquely aesthetic experience, it is that this allegorization of art contextualizes the temporal unfolding of the semantic and visual fields in question to the spatiotemporal dimensions of the world as we know it. This is to say that Butor, in his animation of the journey depicted by Lorrain, allegorizes the creative act not by "giving form to the ideal and the abstract,"[21] as Clifford Gay has defined the function of allegory, and not by encoding in a structure to be deciphered some sort of message separable from the form in which it is conveyed. Rather, Butor's allegorization of art reveals the creative act to be an opening onto the world, in the Heideggerian sense of the term, the perception of the artist, like that of the reader/viewer, constituting a means of integration into it. As Merleau-Ponty has written:

La perception est non perception de *choses* d'abord, mais preception des *éléments* (eau, air . . .) de *rayons du monde*, de choses qui sont des dimensions, qui sont des mondes, je glisse sur ces "éléments" et me voilà dans dans le *monde*, je glisse du "subjectif" à l'Etre.[22]

Our interpretation of *L'Embarquement* as an allegorization of the creative process has focused thus far on the mutual implication of the verbal and the visual narratives. We have seen that the text's positing of the painting as its allegorical other offers proof of the non-representational value Butor assigns to art and allows us to view the Lorrain work as the objective correlative of reality or the world in its function of referential point of departure for all creative production. Butor's text, however, contains within it another text – a series of thirty enigmas that the Queen of Sheba will pose to King Solomon upon her arrival in Jerusalem – and these take up in their own right the question of reference.

Consider, for example, the following excerpt from the seventh text:

. . . tandis que j'ecrivais ce mot, j'ai vu peu à peu mon stylo devenir plus flou. Je l'ai senti encore quelque temps entre mes doigts sans plus le voir, puis même les sensations tactiles, sur lesquelles pourtant je me concentrais, ont disparu l'une après l'autre.

Also, the twelfth text:

. . . un mur de verre devant un mur de carton blanc, devant un mur de poussière, devant un mur de tuyauterie multicolore, devant un mur de paperasses, devant un mur de prétention, d'ignorance, un mur d'insensibilité, d'habitudes, devant un mur de plantes, devant un mur d'eau ruisselante, devant un mur de bruits et de bavardages, de moustaches et de clins d'yeux, devant le mur du son, le mur du sommeil, le mur de la lumière devant la porte entrouverte.[24]

and the thirteenth:

. . . le monde était flou. Lorsque je sortais mon porte-monnaie de ma poche, je n'arrivais jamais à savoir le nombre de pièces qu'il y avait encore à l'intérieur. La marchande de journaux s'amusait à m'en rendre une; la charcutière m'expliquait gentiment qu'il lui en fallait encore. J'avais beau faire mes comptes le soir, cela ne tombait jamais juste. J'avais d'ailleurs souvent de bonnes surprises; les gens ne profitaient pas de mon état.

Mais le lendemain matin, lorsque je reprenais les choses en main, le nombre des pièces ne correspondait plus à celui que j'avais vérifié trois fois la veille. Et pour les numéros de téléphone, c'était pareil; il fallait toujours essayer trois fois d'arriver sur le bon. Et un jour, j'ai dû me rendre à l'évidence: l'écart grandissait . . .[25]

The significance of these enigmas is threefold: First, their subject is precisely that of Butor's work: namely, the conditioning of perception, its relativity, and the contextual factors that determine meaning. Second, the decentering, the undoing that is the focus of each mirrors the deflection of the verbal onto the pictorial meaning of the work that contains them, and vice versa, and is reminiscent of the journey motif in which physical displacement, in this as in so many of Butor's other fictions, is the thematic counterpart of the deconstructive or destabilizing and dematerializing force at play within them. The point is clearly made in each that the assigning of meaning (to text or the objective world) is associative and determined by historical and cultural coding.

Third, the enigmas are significant in their relation to the passage from the Bible cited above. In advancing the narrative line suggested by " . . . and she came to prove [Solomon] with hard questions," Butor's subtexts appear to complete the Biblical passage in much the same way that Butor's *L'Embarquement* does the Lorrain painting. Ironically, however, in their animation of it, they *undo* it for, as enigmas, they deflect meaning – as the "mur de verre" is a throw back onto the "mur de coton blanc," a throw back to the "mur de poussière" and so on – reflecting, once again, the deflection of verbal onto visual signification that is the Butor work. The deflection of the subtext's meaning is of course onto the reader himself, for it is he who, like the "vous" of *La Modification*, must assume the active role in the production of meaning.

We are reminded here of Claude Simon's recent narrative, *L'Invitation* where, despite the official visit of the author and fourteen other intellectual luminaries to the Soviet Union that enframes the fiction, the invitation proves to be not so much Simon's to a foreign land as the reader's to participate in the revitalization of the work that each reading implies. Simon's writing, moreover, pays witness not to the external environs, to the unusual topology that is merely hinted at, but to the internal impressions and fantasies they solicit and, while Simon appears to be inviting the reader into the Soviet Union with him, he is actually turning him away. It is this subversiveness that is striking in both authors and that undermines any attempt on the part of the reader to remain a passive recipient of the fiction at hand.

The reflexivity of *L'Embarquement* is remarkable in that the reflection is not simply to the first or second degree, but infinite. Though not new, for the *mise en abyme* is a hallmark of the Nouveau Roman and can be traced back to Gide and others as well, it is particularly rich in Butor's work where it calls into question not only referentiality but the structure of allegory itself, for it poses, before all else, the question of what may be said to allegorize what. Indeed, we do not and cannot know to what extent the text functions as an allegory of the painting over and above the painting's allegory – in its textually animated form – of the written work. Neither do we nor can we determine which is the proper allegory of the Biblical passage which, in itself, allegorizes what may be said to be another text: the historical event of the Queen's departure. In essence, the *mise en abyme* operating within the text allegorizes that which takes place outside it: The text's specularization of the painting specularizes the Bible specularizing history. And at the center of all this is the enigma, the allegorization of allegory.

Indeed, the enigma is in itself fundamentally allegorical in that it constitutes "a void that 'signifies'," in the words of Walter Benjamin, "'the non-being of what it represents'."[26] "Allegory," to quote de Man, "can only blindly repeat its earlier model, without final understanding . . ."[27] and, hence, it poses a riddle that allegorizes its own allegorical function.

In a recent paper on Duchamp, our colleague, Sidney Feshbach, conspicuously absent from this year's meeting, writes of the artist's *Nude Descending a Staircase # 2* as an allegory of the art world: " . . . it undressed the art world, showed its activities and motions from several angles, saw it descending into the marketplace, revealed its inner mechanisms."[28] Duchamp, Fesbach goes on to explain, "initiated an incisive critique of Modernism as well as the conventions of the bourgeois-capitalist art world."[29] *L'Embarquement de la Reine de Saba* is as playful an allegory of art and the art world as Duchamp's *Nude*. It too shows the "inner mechanisms" of each and narrates the disintegration of the orthodoxies of aesthetics. Like Duchamp, Butor incorporates in his art objects taken for granted in our daily lives and empties tradition – through a privileging of the commonplace – of its hold on art. List after list in *L'Embarquement de la Reine de Saba* of all that is carried by the Queen and her entourage to Jerusalem overcomes the self-reflexivity of the work to re-contextualize it on the order of reference to the cultural norms of a given society. It is precisely this contextualization that invites a reexamination of the kinds of interpretation appropriate to Butor's anaesthetic practice of art.

Montclair State College

NOTES

[1] Gary Shapiro, "Gadamer, Habermas and the Death of Art" in *British Journal of Aesthetics*, Vol. **26**, No. 1 (Winter 1986), p. 39.
[2] Angus Fletcher, *Allegory: The Theory of a Symbolic Mode* (Ithaca: Cornell Univ. Pr., 1964), p. 1.
[3] *Ibid.*
[4] Roland Barthes, *Sur Racine* (Paris: Seuil, 1963), p. 148.
[5] Paul de Man, *Blindness and Insight* (Minneapolis: Univ. of Minnesota Pr.), p. 34.
[6] Carolyn Van Dyke, *The Fiction of Truth: Structures of Meaning in Narrative and Dramatic Allegory* (Ithaca: Cornell Univ. Pr., 1985), p. 27.
[7] *Ibid.* (Underlined by us.)
[8] *Ibid.*
[9] *Ibid.*
[10] Hans-Georg Gadamer, *Truth and Method* (N.Y.: Crossroads, 1975), p. 65.
[11] *Ibid.*, p. 67.
[12] *Ibid.*, p. 67.
[13] Hans-Robert Jauss, cited in de Man, *op. cit.*, p. 174.
[14] de Man, p. 174.
[15] "The First Book of the Kings" in *The Holy Bible*, King James version (N.Y.: American Bible Society), p. 346.
[16] Michel Butor, *L'Embarquement de la Reine de Saba* (Paris: La Difference, 1989), p. 18.
[17] *Ibid.*, p. 27–28.
[18] *Ibid.*, p. 18.
[19] *Ibid.*, p. 20.
[20] *Ibid.*, p. 22.
[21] Clifford Gay, *The Transformation of Allegory* (London: Routledge & Kegan Paul, 1974), p. 28.
[22] Maurice Merleau-Ponty, *Le Visible et l'Invisible* (Paris: Gallimard, 1964), p. 271.
[23] Butor, *op. cit.*, p. 39.
[24] *Ibid.*, p. 53.
[25] *Ibid.*, p. 53.
[26] Walter Benjamin, cited in de Man, *op. cit.*, p. 35.
[27] de Man, p. 186.
[28] Sidney Feshbach, "Marcel Duchamp or Being Taken for a Ride: Duchamp was a Cubist, a Mechanomorphist, a Dadaist, a Surrealist, a Conceptualist, a Modernist, a Post-Modernist – And None of the Above" in: *James Joyce Quarterly*, 26/4 (Summer 1989), p. 548.
[29] *Ibid.*

DAVID K. HECKERL

PAUL DE MAN AND THE QUESTION OF 'DOMINATION FREE' ALLEGORY

I

While talk of an interpretive tradition would be premature, it is nonetheless true that modern assessments of allegory display an intense preoccupation with what Julie Ellison has called the "aggressive tendencies of the mode" (102). For example, the American New Critics John Crowe Ransom and Allen Tate argue that the allegorist's treatment of nature approximates the instrumental, exploitative disposition toward nature expressed in science and technology. Observing that allegory's "primary direction is towards the oversimplification of life which is the mark of the scientific will" (178), Tate suggests that science, despite its scorn for the allegorist's naive belief in the causal efficacy of "magical fictions" (182), has its historical origins in the allegorical desire to dominate nature:

> With the decline then of pure allegory, we see the rise of a new systematic structure of entities called science, which makes good the primitive allegorist's futile claim to the control of nature. Between allegorist and scientist there exists the illusion of fundamental opposition. They are, however, of one origin and purpose. For the apparent hostility of science to the allegorical entities is old age's preoccupation with the follies of its youth. (180)

Tate's essential point is that while a "primitive" allegorist like Edmund Spenser lacked the modern scientist's "systematic method of asserting the will" (183), there remains a fundamental continuity between allegorical poet and scientist in terms of a shared commitment to the power and importance of abstraction. This commitment is predisposed to aggression, however, for abstract thought is "calculated to force the situation upon which it is imposed toward a single direction" (180). If the "situation" in question involves nature, abstraction reduces the rich variety of natural forms and processes to sets of formal, generalizable laws as in theories of gravity or thermodynamics. Admirers of Husserl's exegesis of Renaissance physics in *The Crisis of European Sciences* will recall that Galileo accomplished the conceptual reduction of nature via the imposition of geometrical forms. Geometry, writes Husserl, "aids

M. Kronegger and A-T. Tymieniecka (eds.), Analecta Husserliana XLII, pp. 141–149.

in bringing the sensible surrounding world to univocal determination" (29); in consequence, "nature itself becomes . . . a mathematical manifold" (23).

For Tate, then, there is no substantive distinction between Galilean physics, which aggressively seeks the "univocal determination" of nature in geometry, and the allegorical method used by Spenser to construct *The Faerie Queene*. Ultimately, it matters little that Spenser, the poet, presents his abstractions in the guise of sensuous landscapes and complex personifications while Galileo, the scientist, presents his abstractions in comparatively spare form as geometrical figures. In both instances, the particularities of nature and experience are sacrificed in deference to abstract configurations or didactic themes. In Galileo's system, the sensuous materiality of living nature is, in the last analysis, always reducible to geometrical forms and mathematical formulae. And despite the vibrant, detailed presentation of character, event, and landscape in *The Faerie Queene*, it remains the case that each concrete image is related in some way to the abstract, allegorical clash between Virtue and Vice.

As John Crowe Ransom has written, allegory is best defined as a "discourse in things, but on the understanding that they are translatable at every point into ideas" (874). It is precisely this reduction of "things" to "ideas", however, that serves to qualify allegory, not as poetry, but as science. For Ransom, allegory is "an exercise which is really science but masquerades as poetry by affecting a concern for physical objects" (874). The implication here is that, in contrast to allegory, authentic poetry is sincerely interested in the sensuous appearance of the things themselves. Authentic poetry has no "wish to impair the thinginess of things" (874); indeed, the "art of poetry" depends on the "faculty of presenting images so whole and clean that they resist the catalysis of thought" (873). Allegory, which Ransom describes as "militant, always sciencing and devouring" (876) is diametrically opposed to the true poet's "aspiration towards innocence" which humbly seeks "to respect and know the object as it might of its own accord reveal itself" (876). The allegorist, claims Ransom, assumes that the "things as things do not necessarily interest us" (875); rather, allegory ministers to our aggressive need to "view the world under universal or scientific ideas . . . and this is because the ideas seem to make not for righteousness but for mastery" (875). In exalting the abstract idea at the expense of concrete things, allegory "symbolizes the predatory method, like a sawmill which intends to reduce all the trees to fixed unit timbers" (877).

Ransom obviously shares Tate's sense of allegory as destructive of nature, and this is explicitly linked, as in Tate's analysis, with the aggressive process of abstraction. In a passage that vividly evokes the violence of this process, Ransom compares the activity of the scientist, and by extension that of the allegorist, to a hunting expedition in which the concrete image or thing is transfixed by abstract "values" and thereby destroyed:

> It is not by refutation but by abstraction that science destroys the image. It means to get its "value" out of the image, and we may be sure that it has no use for the image in its original state of freedom. People who are engrossed with their pet "values" become habitual killers. Their game is the images, or the things, and they acquire the ability to shoot them as far off as they can be seen, and do. (872)

"It is thus," Ransom concludes, "that we lose the power of imagination, or whatever faculty it is by which we are able to contemplate things as they are in their rich and contingent materiality" (872). It is precisely because allegory abjures this humble, receptive disposition toward nature in favor of an aggressive will to power that it fails as poetry. Allegory, Ransom notes, "tries as hard as it can to look like physical poetry, as if it proposed to conceal its medicine, which is the idea to be propagated, within the sugar candy of objectivity" (874). However, in adhering to Ransom's metaphor, we can only say that the "sugar candy of objectivity" dissolves to nothing under the aggressive heat of abstraction.

II

In a recent essay appropriately entitled "Aggressive Allegory", Julie Ellison reiterates for a postmodern audience the themes and concerns addressed by Tate and Ransom. For example, Ellison follows her predecessors in locating "allegory's violent reductiveness" (103) in the "link between abstraction and violence" (102). In a heightened rhetoric similar to Ransom's, Ellison speaks of the "murderous consequences of allegory for the object" (103). Ellison departs from Ransom and Tate, however, in her distinctly postmodern understanding of allegorical abstraction. For Ellison, allegories are characterized by conflicting sets of "binary oppositions" (101) tailored to "represent a relationship of oppositional difference" (101). Taking Ralph Waldo Emerson's prose as her example,

Ellison describes the conflict between "idealism, mind, soul, intellect, Reason . . . [and] their opposites; materialism, body, nature, understanding" (102) as essentially allegorical:

> Emerson does not write allegories, but uses allegory as a tactic . . . Terms like idealism and materialism, soul and nature, poet and skeptic, behave like figures borrowed from allegory. They are images of virtues, vices, or other invisible attributes engaged in actions, primarily contests and conquests. (102)

Ellison's claim, an important one, is that definitions of allegory need not be confined to strictly literary modes, i.e., to medieval morality plays and works like *The Faerie Queene* or Dante's *Divine Comedy*. On the contrary, the term 'allegory' can be accurately applied to any cognitive or representational system that organizes experience in terms of abstract, binary oppositions. Thus, in the wake of postmodern deconstructions of such systems, it becomes possible to read the course of Western thought as a veritable litany of allegories: speech and writing, grammar and rhetoric, subject and object, structure and genesis, signifier and signified, masculine and feminine, nature and culture . . . all of these binary pairs indicate, in Ellison's view, a pervasive tendency toward allegorical thinking. As Ellison observes, the "reductions such namings entail can usefully be thought of as allegorical" (102).

The claim that binary oppositions are necessarily reductive returns us to the theme of aggressive allegory. Ellison, in common with many postmodernists, maintains that the classifications and categories used to describe experience can always be unmasked as arbitrary impositions of power; indeed, deconstructive approaches to these categorizings routinely attempt to recover an excluded term or concept that, when re-introduced into the binary relationship, effectively collapses the distinction. As mentioned, what informs these deconstructive strategies is the belief that relations of difference are always reducible to questions of power. Obviously, this belief entails the claim that descriptions of categorical differences do not refer to objectively existing states of affairs. This denial of reference is crucial, for if the differences in question *did* in fact exist independently of their descriptions, then such descriptions could not be reduced to acts of aggression. On the contrary, such descriptions would be seen as referring to what is actually the case. We would do well to keep in mind, then, that Ellison's reduction of allegorical thought to aggression presupposes the impossibility of accurate reference. The allegorical construction of difference is aggressive only on condition that the differences described do not exist independently of their descriptions. As will be seen, however, Ellison is far from secure in her reduction of allegory to aggression.

III

At a critical moment in her essay, Ellison is obliged to dismiss Paul de Man's theory of allegory precisely because he "denies [its] aggressiveness [in] distinguishing the will to difference from the will to power" (104). De Man, Ellison notes, "recommends" allegory as providing "sophisticated knowledge of the difference between man and nature" (104) in stressing the ontological distinction between linguistic sign and natural object. In claiming, then, that authentic knowledge of this distinction is possible, de Man implies that the abstract, 'allegorical' distinction between nature and culture is *not* reducible to questions of power and aggression. Not surprisingly, Ellison construes de Man's "curiously benign" (104) view of allegory as suppressing the connection between allegorical thought and "violence toward nature" (104). "By divorcing reflection and thus difference from questions of power," de Man hopes to "sanitize" allegorical thought (104). Indeed, de Man "wishes to make his own critical allegorizing nonviolent" (114). Predictably, Ellison concludes her engagement with de Man by stating that "he does not convincingly lay to rest the connection between allegory and aggression" (104). Ellison is allowed this hasty dismissal, however, only because she avoids confronting de Man's arguments in defense of the ontological distinction between linguistic and natural forms. The remainder of this essay is devoted to assessing the plausibility of de Man's position in view of his implicit claims for a 'domination free' allegory.

IV

In his influential essay "The Rhetoric of Temporality", de Man maintains that allegory illuminates the "disjunction between the way in which the world appears in reality and the way it appears in language" (191). More specifically, illumination occurs when the allegorist reflects on the purely intentional status of his own constructs, and of linguistic constructs in general. Thus, it is by reflecting on the role of intentionality in the constitution of those objects we call 'texts' that we access the radical difference between linguistic signs and natural entities.

Interestingly, de Man's major reflection on intentionality transpires in context of the critical position whose main practitioners included John Crowe Ransom and Allen Tate. In "Form and Intent in the American New Criticism", de Man enquires into the reasons behind the vigorous "rejection of intentionality" (25) expressed in Wimsatt and Beardsley's famous notion of the "intentional fallacy". After a lengthy and intricate analysis, de Man concludes that the New Critics spurned intention because the "introduction of the principle of intentionality would imperil the organic analogy" (28) on which the New Critics based their understanding of literary form. Inherited from the romantic tradition of organicist aesthetics as developed by Kant and Coleridge, the "organic analogy" expressed the radical autonomy of literary forms as spontaneously developing, self-contained unities. To propose, as de Man ultimately does, that literary form emerges through the intentional activity of the creative artist would immediately compromise the autonomy of the literary object, thereby removing it from the realm of nature.

De Man correctly recognizes that what distinguishes natural phenomena such as weather, daffodils, and nightingales is that their material being does not depend on the intentional activity of human agents. It would be absurd to claim, for example, that Newton's cognition of gravitational law caused the appearance of tidal movements on the ocean's surface. Tidal movements caused by the gravitational force of the moon would occur regardless of whether or not Newton was present to observe them, and we can say the same about all non-human, natural phenomena. On the other hand, the existence of educational systems, computers, airplanes, university buildings, literary forms – the totality of what we commonly call 'artefacts' – is entirely dependent on the intentional and/or physical activity of human agents. Roman Ingarden, whose work in phenomenological aesthetics provides the foundation for de Man's meditations on intentionality, describes the mode of being of all artefacts as follows:

> . . . purely intentional objects are not self-existent. They are objects which derive their existence and their entire endowment from an intending experience of consciousness (an "act") . . . They would not exist at all but for the performance of acts of this kind; yet, because such acts are performed, these objects do exist, but not autonomously. (47)

There is little question that artefacts exist, but as Ingarden points out, their existence presupposes the intentional and/or physical activity of human beings; an artefact's ontological condition of possibility resides, not in itself, but in the prior existence and consequent execution of intentional activity. Thus, as de Man writes, the "difference between the stone and the chair distinguishes a natural object from an intentional object. The intentional object requires a reference to a specific act as constitutive of its mode of being" ("Form" 24). Whereas the ontology of literary forms can be properly cognized only by first acknowledging their causal dependence on the activity of human agents, the natural entity remains authentically autonomous in that its material origination does not depend on human intentionality.

V

The fateful question now presents itself. Does de Man's description of the difference between intentional and natural forms exemplify an instance of binary, allegorical thinking that is 'domination free'? Recalling Julie Ellison's comments, what makes de Man's thought allegorical is that it cleaves to the abstract, binary distinction between Nature and Culture by separating, on ontological grounds, those things we call 'literary forms' from the things we call 'natural objects'. We need to ask, then, whether de Man's categorizations are aggressive, arbitrary impositions of power, or whether, on the contrary, such categorizations provide epistemic access to objectively existing states of affairs. The best way to determine an answer, it seems to me, is by subjecting de Man's distinction between intentional and natural forms to a vigorous deconstruction. If de Man's categories come through such a procedure unscathed, we have good reason to believe that his descriptions have epistemic validity, and that they are therefore irreducible to questions of aggression and power.

Applied to de Man's distinction between linguistic and natural forms, a deconstructive strategy can attempt to collapse the distinction in one of two ways: 1) it can claim that intentionality plays no constitutive role in the material emergence of linguistic forms, in which case such forms can be said to possess the radical autonomy of natural entities. By thus banishing intentionality, the distinction between linguistic and natural forms collapses. 2) it can claim that natural phenomena, far from existing independently of culture, are actually constituted by cultural activity. In this case, the distinction between literary and natural forms is collapsed by proclaiming that all the world is an artefact. Both strategies are provocative, but are the resulting world scenarios plausible? In the case of strategy 1, banishing notions of intentionality, subjectivity and agency projects a world in which literary forms are self-constituting entities, emerging, as it were, of their own volition. That this view constitutes a rather severe case of reification should be obvious, for it is simply false to maintain that linguistic forms organize themselves independently of the intentions, plans, and desires of human agents. As for strategy 2, the reduction of natural phenomena to the status of artefacts projects a causal absurdity which, as illustrated above with respect to Newton, is simply incredible. Cultural representations of the extinction of animal species cannot be said to have caused the extinction of such species. It is certainly true that this state of affairs receives expression in cultural discourse, but admitting nature's *semantic dependence* on culture does not erase its *causal independence* of culture.

It is hoped that this brief account of the implausible ontological scenarios generated by attempts to deconstruct the distinctions between linguistic and natural forms will have confirmed the epistemic validity of de Man's allegorical thought. Unless we are willing to seriously entertain the belief that literary forms are self-organizing, or that cultural activity causes or originates natural phenomena, we must concede the validity of de Man's ontological distinctions. Far from being an aggressive, arbitrary imposition of abstractions, de Man's description of the opposition between intentional and natural forms provides epistemic access to what is in fact the case independently of such descriptions. Linguistic forms, like all artefacts, really are dependent on the intentional activity of human agents, just as natural forms really do exist independently of cultural activity. In successfully stabilizing this distinction, de Man demonstrates that not all abstractions, such as that between Nature and Culture, are reducible to acts of aggression. In de Man's work, we have evidence of 'domination free' allegory.[1]

McGill University

NOTE

[1] Those readers convinced of de Man's canonical status as a 'deconstructive' critic will undoubtedly condemn as ludicrous and heretical my account of de Man as an ontological realist. My reply to this objection is that some recent commentary on post-

structuralist thought has stressed significant differences between de Man and Jacques Derrida, and that the basis of these differences lies, as Richard Rorty has observed, in de Man's serious commitment to a phenomenological ontology:

In particular, Derrida never takes the metaphysical distinction between what [Umberto] Eco calls 'the universe of semiosis' and some other universe – between culture and nature – as seriously as de Man did. De Man makes heavy use of the standard Diltheyan distinction between 'intentional objects' and 'natural objects.' He insists on contrasting language and its imminent threat of incoherence . . . with the putatively coherent and unthreatened rocks and quarks. Derrida . . . edges away from these distinctions, viewing them as just more remnants of the Western metaphysical tradition. De Man, on the other hand, makes them basic to his account of reading. (101–2)

I agree. See Richard Rorty, "The Pragmatist's Progress," *Interpretation and overinterpretation*, ed. Stefan Collini (Cambridge: Cambridge UP, 1992), 89–108.

REFERENCES

de Man, Paul, "Form and Intent in the American New Criticism," *Blindness and Insight* (Minneapolis; University of Minnesota Press, 1983), pp. 20–35.

de Man, Paul, "The Rhetoric of Temporality," *Blindness and Insight* (Minneapolis: University of Minnesota Press, 1983), pp. 187–228.

Ellison, Julie, "Aggressive Allegory," *Raritan* (Winter 1984) pp. 100–15.

Husserl, Edmund, *The Crisis of European Sciences and Transcendental Phenomenology*. Trans. David Carr (Evanston: Northwestern UP, 1970).

Ingarden, Roman, *Time and Modes of Being*. Trans. Helen R. Michjeda (Springfield: Charles C. Thomas, 1964).

Ransom, John Crowe, "Poetry: A Note in Ontology," *Critical Theory Since Plato*. Ed. Hazard Adams (New York: Harcourt, 1971), pp. 871–81.

Rorty, Richard, "The Pragmatist's Progress," *Interpretation and overinterpretation*. Ed. Stefan Collini (Cambridge: Cambridge UP, 1992), pp. 89–108.

Tate, Allen, "Three Kinds of Poetry." *Essays of Four Decades* (Chicago: Swallow Press, 1968), pp. 173–97.

NANCY CAMPI DE CASTRO

ALLEGORY, YOU ARE WOMAN

It is our intention to start our study by presenting a modern fable which was written by Guillermo Cabrera Infante, a contemporaneous Cuban writer. Named "The Fox and the Grapes", this is his text:

A fox was hungry and as she was unusually vegetarian (there is nothing a fox wouldn't make to be up to date), she looked at some grapes which were near her, but above her. She jumped once, twice, and one more time, but she couldn't reach them. The fox looked at the grapes for the last time and looking carefully (it's astonishing how things are carefully seen when for the last time) exclaimed: "It doesn't matter: I don't want them: they are green." A crow, which was near by on the way to another fable, looked at the fox and then to the grapes and said: "It isn't strange: the grapes look green because they are green grapes." The fox didn't answer, perhaps she didn't listen, perhaps she was proud, but undoubtedly because foxes cannot talk to crows. *Moral*: A fox is an animal which hasn't the gift of speech, but on the other hand, may suffer from daltonism.

Fable is a form of allegory; it is a developed allegory in which an idea is intended to be expressed through elements that are substantiated in the characters of this kind of narrative. A fable presents in its structure all elements which are essential for allegory: it is a concrete representation of an idea; each one of its allegorical components has a meaning other than its own meaning, and not only what may be understood primarily; there is a link between what is shown and the underlying meaning.

We know quite well the original text of the fable "Sour Grapes" which is typically allegorical: a fox, symbol of astuteness, being ever hungry was quite close to what would satiate her hunger, that is, grapes; since she could not reach the grapes, she resumes her astute attitude and she disdains the object of her delight. Astuteness wins every time – this is the allegorical reading.

Guillermo Cabrera Infante reclaims the fable, reclaims the allegory and enlarges its meanings. Cabrera Infante makes use of all his modernity: he breaks, shatters allegorical conventionality – the astute fox, who does not want the grapes, for they are green, finds her stock of astuteness diminished by the crow's diagnosis: the grapes look green, for they are green, grapes. As another element is inserted – the crow,

M. Kronegger and A-T. Tymieniecka (eds.), Analecta Husserliana XLII, pp. 151–157.

"on the way to another fable" – an objective rationalization is offered, and it introduces another allegorical reading: there is something wrong with the fox. The fox is hungry, the grapes are not "green", and the fox gives them up. Is she foolish?

The crow, who, out of vanity, had already been cheated by the fox in another fable – he let the cheese desired by the fox fall from his beak – now recovers his dignity and makes a clear and rational analysis: "the grapes look green for they are green grapes." The fox is foolish. And doubly foolish, because if she did not know beforehand that the grapes were green grapes, now she knows it through the speech of the crow. But the fox does not react, she does not defend her astuteness. And then, at its end, the fable starts to also recompose the traditional profile of the fox, when it explains: "foxes cannot talk to crows . . . ". And in the moral, at the allegorical end of the fable, astuteness is partially given back to the fox, who is nevertheless, according to Cabrera Infante's narrative, an astute animal. But she is an animal who has limitations, who doesn't have the gift of speech, and therefore, who does not know how to use other codes to communicate through speech, and who, besides, suffers from daltonism: the fox keeps her astuteness, but she has deficiencies, even a visual deficiency.

Another reading for meaning-enlargement which may be introduced concerns the crow: is he, as a prototype of seriousness, telling the truth? Or in his strange presence in this fable, is there only a malignant intention of personal revenge against the fox, giving her a verdict which steals part of her astuteness? Many other questions may follow these. The truth of this fable, or of this allegory, must be dug out by the reader, and each one will find his/her truth, according to his/her reading of the text.

What is meant to be emphasized is that everything functions allegorically to enlarge the meaning of the fable. Closed and ready conventionality, sanctified by culture through the time, is questioned: astuteness exists, but the triumph of its agent is less crushing; its agent is not a winner without faults, he/she is not able to defeat everybody. The truth is questioned: what is the reader's guarantee? Should he believe the crow's truth, the fox's truth, or the underlying truth of the moral of the fable?

This introduction intends to demonstrate that allegory, in modern times, shows itself through multiple faces. From its genesis as a participant in rhetoric until the present time, allegory has been creatively worked. One of the most significant aspects of allegory in modern times is that it deals with the main point: the relationship between a work of art and how it is interpreted by the spectator.

A work of art is a semantic act, for it makes a significant mark in the world. A work of art tries to report reality and reality tries to report itself through the work of art. According to the etymology of the word allegory, "to say the other", a work of art says its other and tells itself through the other, as an authentic allegorical process. This is further reinforced, emphasized and reiterated within contemporary works of art.

Even keeping in mind that allegory, through time, is a component of culture and that at times it has been useful to legitimize interests, especially those which represent power and domination, and that allegory thus becomes conventional by a continual and ceaseless repetition of the same signifier for the same significance, it is necessary to recover by other means the inherent idea of allegory which helps to transcend conventions. Mechanical acceptance of allegory, without questioning it, leads to ideological domination. Synchronic, in many cases, allegory has been the paradise of conservativeness and immobilism, which gives it an authoritarian character as a progress inhibitor. Allegory has been used on societies in different eras as a strategic ideological instrument.

It is necessary to introduce an innovative inquietude into allegory: thus, it will show itself in a renewed and rich expressive form. This happens when it is possible to unleash meanings in allegory, and thus what seemed to be old becomes new. This is what we call allegorization of allegory, that is, an allegorical reading of allegory itself. It is essential to reach the other side of the mirror: to go beyond the specularity which itself constitutes a work of art. So the nature of allegory must be examined as construction. And it must be deconstructed, even in a playful way, in order to be known and revitalized – this is what is seen in contemporary – works of art.

We live in an era of change, an era of great velocity. Human life and human art express it. Contemporaneity in art, or postmodernism, is conspicuous for its contradictory nature which is characterized by ambiguity and subversiveness. The result of this game of oppositions is the paradoxal performance of works of art, which are stretched between extremes.

The value of the role of spectator is increased: he/she is not passive anymore – or the "solo receptor", according to Julio Cortázar. The spectator acts, interferes, creates. His/her performance, unlike those of other epochs, is no longer an "innocent" performance without intentions which, in terms of a work of art, approaches a "zero point"; the spectator, nowadays, interacts with the work of art, he/she is attracted to it by its playful texture which has no intention of catching his/her attention, but rather of drawing him/her into the game of art, as one of the releasing processes of man's plenipotence. Then the spectator starts interpreting the work of art, as well as his/her own life.

But contemporary man, as a contradictory and postmodern human being, is surrounded by allegories which try to hide the contradictions from which they result. The allegory of woman is one of the biggest allegories of the contemporary era. According to Jacques Derrida, woman is the most complete allegory of truth in Western culture and femininity is conceived as the essence or the truth of woman. To simplify Derrida's affirmation as far as possible and to try to exemplify it, let us say that the increasing tendency in the Western world to fetishize woman is remarkable. The more closed and authoritarian a society is as a source of culture, the greater the process of fetishization. Woman assumes the character of an allegory of femininity, with her standard body and her standardized beauty. These are enforced patterns within essentially allegorical cultural structures, but they are also conventional. What is apparent is that a search for a semantic alterity propelled by reality, one which would result in authentic allegory, an allegory with a different meaning and opposed to convention, does not take place for woman.

Historically, two labels have covered woman like two masks: femininity and feminism.

Femininity is the oldest mask, that closest to woman. Woman in man's world is the allegory of femininity: she is the personification of femininity, which is opposed to masculinity. Woman must be feminine in order to attract the male, to be desired by him. Mozart's Don Giovanni has already evidenced this male need for woman: "Lasciar le donne! Pazzo! Sai ch'elle per me son necessarie più del pan che mangio, più dell'aria che spiro!"

Femininity results in a wish to give voice to sexual difference, establishing the position of woman-object as a homogeneous significance, as a conventional and closed allegory.

And what is woman's posture on this femininity in the contemporary world? Essentially there are three positions in the Western world which synthesize women's attitudes: there is the fight for equal rights, representing equal access to the symbolic order; there is the rejection of the masculine symbolic order for the sake of change, and there is also a rejection of the dichotomy of masculine and feminine. The first position, which aims at equal access to the symbolic order, is taken by Anglo-American women theorists, and it intends to signify a struggle against patriarchy; the second position, that of rejection of the masculine symbolic order, which is taken by French women theorists, also represents an opposition to patriarchy, but it tries to emphasize the unique aspect of femininity; the third position is taken by American women theorists and results in a critical psychoanalytic-feminine view which considers human identity as a cultural construction that is evidently problematic for both men and women.

Nowadays, it is inferred that what is called femininity was determined by the discourse of patriarchy, which was involved in binary structures that became rigid as though these oppositions had always existed, allowing neither possible questions nor analyses concerning their nature.

The allegory of femininity, represented by woman, produced in this century a "sui generis" movement: feminism, a socio-political movement that emerged by the end of the sixties. It opposes the patriarchal vision of society and sexism, and evidences a pluralistic tendency designed to attain the widest possible variety of results. The fiercest fight of feminism is against the domination of women by males as the most fundamental concept of power and the most piercing ideology of Western culture. Everything would appear to be contaminated by this patriarchal ideology: Simone de Beauvoir, for instance, when she wrote *Le deuxième sexe*, would seem to have been deeply influenced by Sartre's phallocentric categories. Feminism, as a presence in our world, has so far not annulled the conventional allegory of femininity. It is inferred that femininity is a cultural construction expressed through enforced patterns that, however, must appear to be natural. There would seem to be a female essence named femininity: this is affirmed by the culture of patriarchal society which deals with static binary oppositions such as male and female, feminine and masculine, man and woman, leading to many others such as activity and passivity, sun and moon, night and day, father and mother, and intelligence and sensibility, all of which make up a patriarchal system of principles. When psychoanalytically analyzed, each opposition works as a hierarchy in which the female extreme always represents the weakest or the most negative instance. There are theorists who conclude that any common pattern of femininity found will only be the result of a long history of oppression.

Julia Kristeva, for example, in her position on femininity, defines women as being marginal to the symbolic order and to society, and she conceives of femininity as denial, absence of significance, irrationality, darkness, chaos and, finally, as not-being.

Therefore, what is the true essence of woman? Is there indeed such an essence, which characterizes only woman? Or does this essence merely reveal and emphasize the already standardized opposition of non-femininity, or masculinity?

One question raises others: if this femininity exists, will it be different from masculinity, considering the snare of binary opposition? Further: will all women be feminine? Or will there be "feminine" men? Is femininity a theoretical-cultural construction? If so, would it be a characteristic of a determinate sex? Or could we speak of psychoanalytical elements that characterize human beings – men or women – implying that there are particular ways of assuming life and of transforming it?

The search for the answers to these questions leads to the necessary allegorization of the allegory of femininity. A new and contemporary reading of woman within the Western world will reveal a further feeling: it will be an allegorical reading that points to alterity, which reflects upon itself, which devotes itself to an exercise of freedom.

In the same way, it is possible to speak of allegory in the contemporary world and in contemporary art: if postmodern art is conceived as a social sign which inevitably emerges from other signs that belong to systems of signification and value, it is time to create new signs which will bring new meanings even to ancient allegories.

Thus we would be able to exhort allegory, personifying it in terms of art and life: "Allegory, you are woman: fight against conventions, exhibit yourself, reveal your contradictions – this is the only way to revitalize yourself, by renewal through creativeness and freedom and the subversion of fixed schemes; this is the only way to state the truth and to make the truth known."

We conclude our study with the desire that man and woman may speak for mankind in allegory and in life, conferring, in art and in life, new shapes to this mankind which is still so fugitive, so enigmatic, but in essence so powerfully creative.

Universidade Federal de Juiz de Fora, Brasil

REFERENCES

Cabrera Infante, Guillermo, *Exorcismos de esti(l)o* (Barcelona: Seix Barral, 1976).
Castro, Nancy Campi de. "O desempenho do leitor contemporâneo." *Revista de cultura Vozes* (Petrópolis) V. **78**, n. 7 (Sept. 1984), pp. 25–32.
Cortazar, Julio, *Valise de Cronópio* (São Paulo: Perspectiva, 1974).
Derrida, Jacques. "Choreographies: An Interview with Jacques Derrida." Edited and translated by Christie V. McDonald. *Diacritics*, V. **12**, n. 2 (Summer 1982) pp. 66–76.
Kristeva, Julia. *Histoires d'amour* (Paris: Denoël, 1983).

PART III

A group of participants at the Luxembourg Congress. From left to right: Ana-Maria Holzbacher, Margarita Amieiro Alfaro, Cynthia Ruoff, Marlies Kronegger, Colette Michael, Pascal Voilley, Pilar Suárez Pascual.

PILAR SUÁREZ PASCUAL ET MARGARITA ALFARO AMIEIRO

L'ALLÉGORIE DU MAL. DEUX ÉPOQUES, DEUX PERSPECTIVES

INTRODUCTION

Les images littéraires constituées à partir de l'idée du mal ont toujours été présentes dans la littérature universelle, en donnant naissance à d'innombrables figures rhétoriques de sens.[1] L'intérêt littéraire de ce concept nous a poussées à essayer une approche sur le processus allégorique que celui-çi subit dans deux domaines tout à fait différents; le genre dramatique des *Mystères* et *La Fin de Satan* de Victor Hugo. Nous avons justifié cette étude comparative dans deux manifestations si disparates, en nous appuyant sur l'identité de la matière thèmatique qui informe ces oeuvres: le mythe biblique.

Notre intention est de mettre en parallèle le traitement, la fonctionalitè et la signification qu'un même topique littéraire peut avoir dans deux processus d'allégorisation qui présentent une forme et une signification différentes étant donnè la nature de chacune des oeuvres. D'ailleurs, aussi bien dans un cas comme dans l'autre, ces ouvrages intègrent la figure du mal dans un texte qui essaie de présenter un modèle cosmologique propre. Cela nous a fourni le point de départ pour considérer la possibilitè d'entreprendre ce type d'étude.

Afin de résoudre le problème de travailler sur deux genres nous avons réfléchi sur le concept d'ethos tel qu'il est conçu par le Groupe M[2] – "situation affective suscitée dans le récepteur par un message dont la qualité spécifique varie en fonction d'un certain nombre de paramètres" – et à partir de cette idée nous avons articulé notre analyse autour de deux points fondamentaux: le contexte intra-diégétique déterminé par la propre nature des oeuvres, et les circonstances d'émission et de réception qui déterminent leurs respectives présentations et qui conditionnent les valeurs significatives du processus alléqorique, et les valeurs significatives qui s'en dégagent.

M. Kronegger and A-T. Tymieniecka (eds.), Analecta Husserliana XLII, pp. 161–183.

I. LA PRÉSENCE DU MAL DANS LES MYSTÈRES

Les circonstances d'émission et de réception

La nature du Mystère. À la fin du Moyen Âge le diable devient l'une des unités culturelles[3] les plus obsédantes de l'époque; en gardant les valeurs qu'il possédait dans le récit biblique, sa figure éprouve un essor à cause de la préoccupation des hommes du XVème siècle pour la mort, la damnation et le monstrueux. Ils essaient de s'en défendre de différentes façons – il faut se rappeler l'abondance des procès contre les sorciers – et la littérature se constituera en moyen pour le faire.

Le Mystère est l'une des minifestations privilégiées pour pouvoir étudier ce phénomène, non seulement parce que sa nature thématique demands la présence du démon, mais parce que ce genre se fonde précisément sur la mise en scène de toutes les valeurs culturelles – positives et négatives – qui articulent la civilisation urbaine de l'époque.

Ce phénomène théâtral naît avec la bougeoisie et son apogée; c'est la forme d'expression de son pouvoir qui s'est prolongé vers d'autres activités du monde urbain, et très religieuse acquiert, donc, le caractère d'une allégorie du pouvoir à travers de deux éléments très importants:

- la réduction progressive de la communauté;
- l'appropriation de l'espace qui représente la totalitè de la ville: la place publique.

L'homme de la cité médiévale ne se définit pas seulement par sa profession, sa condition ou sa fortune; mais par l'appartenance à un groupe social; il est toujours intégré dans une société qui lui procure du secours. C'est le concept de "clan"[5] qui domine la vie sociale de cette époque, très spécialement en ce qui concerne les aspects de groupe homogène, exclusif et restreint qui se place sous un élément emblématique qui est la figure du Christ et de l'Église.

Tenant compte de cela, il faudrait distinguer entre le mythe biblique en soi, qui fournit la matière thématique pour les Mystères, et le Mystère en soi. Celui-ci est une allégorie car il devient une traduction conceptuelle de l'hypotexte: c'est-à-dire ce qui se joue n'est pas seulement l'histoire de la Rédemption du monde – ce qui appartiendrait plutôt au domaine du Jeu Liturgique –, mais cette histoire est reduite à l'établissement des paramètres qui animent la communauté urbaine de l'époque. C'est la ville qui s'y voit refletée, c'est son triomphe qui est célébré: la différence fondamentable c'est qu'il y a eu une appropiation d'un espace et d'uin mythe autrefois universel, vers un cadre restreint. Le Triomphe du Christ deviendra celui de la ville. L'élément social a rejoint le religieux, se trouvant ainsi investi d'un aspect transcendant et spirituel. C'est pourquoi il faut interpréter ces représentations en parallèle: en gardant leur sens litéral, mais tenant compte de la signification que les émetteurs et les récepteurs accordent au message en fonction d'un univers de référence commun.

L'Autre et le danger. Au XVème siècle la ville française subit des guerres, des famines, des morts; tout cela crée un sentiment d'instabilité. A cela s'ajoute la peur qu'elle a des éléments hostiles étrangers à la cité – révoltes paysannes, bandes de marginaux qui attaquaient les gens. Ces élements cristallisent la peur de la société. Ils représentent les possibles changements adverses de la fortune, et sont chargés de tous les crimes imaginables. Cette conception de l'autre nous fait supposer que les gens de cette époque donnent forme à leurs peurs et les actualisent dans des images appréhensibles à travers les sens: c'est le diable, c'est le juif, c'est le monstre.

La diablerie est l'une des manifestations dramatiques qui répond à ce besoin; et c'est dans ce contexte infra-théâtral que le personnage du Démon se développe.

La mise en scène de la déchéance. La naissance du mal.

Le premier fatiste à accorder une plus grande attention à la figure du Diable c'est Arnoul Gréban,[8] dans le *Mistere de la Passion de Nostre Seigneur.*[9] Il encadre le drame de la Rédemption dans l'ensemble d'une dégradation préalable de la création due à la naissance de l'esprit du mal. Celle-ci est présentée à partir d'une attitude de révolte de la part de Lucifer, qui provoque la distribution de deux espaces opposés et irréconciliables et qui sont les résultats d'un processus diaïrétique entrepris par le héros de Dieu – Saint Michel – et dans lequel lucifer est le grand vaincu et l'éternel damné.

L'allégorie du Diable commence à se dessiner à travers la naissance de son espace. Ce dernier processus est articulé à partir de trois métaphores fondamentales:

1. La chute, présentée à travers la précipitation des diables dans les enfers vv. 178–188.

Cette action établit la division définitive des espaces transcendents, en donnant raison de la catégorisation du cosmos médiéval en forces positives et négatives. Les diables deviennent des personnages immobiles, à cause de leur impossibilité de s'améliorer et de s'évader à leur nouvelle ubication:

Lucifer:

"Hélas
Or n'est espoir d'en saillir Hors" (vv. 211–212)

2. Les démons apparaissent aussi comme des personnages plongés dans le chaos. Cela s'exprime à travers la réalisation d'une descente qui renforce le processus de dégradation éprouvé par les démons qui s'éloignent de leur statut précédent d'habitants du Paradis.

D'ailleurs l'Enfer se constitue en espace nettement marqué par l'absence d'harmonie, conséquence du désespoir éprouvé par les diables quand ils sont conscients de leur déchéance. Ce désespoir est exprimé à travers le bruit: les cris, les imprécations créent un climat aussi bien comique que tragique, car ils se révèlent en personnages impuissants qui déplorent leur nouvelle condition à travers l'agressivité, exercée parfois contre eux-mêmes.

"Yci doivent fere grant tempeste" (int. vv. 228–229).

3. La déchéance est aussi renforcée par un processus scénique: la transmutation physique à travers laquelle l'ange devient bête.

"Ycy cheent et tresbuchent/ les mauvais anges en enfer en fourme de deables" (int. vv. 204–205).

Le mal a été visualisé et raisonné à travers un systéme de représentation sensualiste sous les marques sémantiques de l'animalité, l'abîme et l'absence d'harmonie.

La projection du Diable

C'est en fonction de l'existence du personnage du diable et de sa projection actantielle sur la terre que la rupture du contrat entre Dieu et les hommes est expliquée. Nous nous trouvons devant le deuxième point de ce processus allégorique: le cosmos est conçu comme un espace bipartite dans lequel se déroule un tour de forces qui atteint son plus haut degré d'expression quand les Diables doivent affronter l'incarnation du principe positif: le personnage du Christ.

L'auteur qui a le mieux exprimé ce dernier conflit a été Jehan Michel dans *Le Mistere de la Passion de Nostre Seigneur JesChrist.*[11] Nous voyons dans cette pièce comment, afin d'éliminer leur nouvel ennemi, les diables, à travers Sathan, agissent encore une fois sur la terre et ont recours à un adjuvant: le Juif.

Ce personnage devient une autre métaphore du mal établie en fonction de sa condition de déicide et de son refus de la figure du Christ. Sa différence fonctionnelle postérieure par rapport aux Diables – c'est lui qui finalement va procurer le triomphe du héros malgré les essais des Démons d'arrêter cette action – ne l'empêche pas de devenir un personnage qui participe de la nature démoniaque à cause de sa mécréance et de l'intentionnalité négative qui guide son actuation, pourtant positive.

C'est précisément du point de vue des besoins créateurs que ce personnage peut être considéré isomorphe des Démons, car son existence répond aussi au besoin de personnifier le mal, cette fois dans un type humain.

Pour comprendre l'action d'auto-abolition entreprise par les Enfers il faut tenir compte que l'action des Mystères est fondée sur une double isotopie: la première – ou isotopie superficielle – marque la dégradation physique du héros, tandis que l'isotopie profonde montre le processus de son triomphe. La structure actantielle du mythe christologique est articulée sur une structure synthétique: c'est à dire que la dégradation éventuelle de Jésus est le pas préalable pour une amélioration définitive. Les démons sont tombés dans leur propre piège, car quand ils s'acharnent à détruire le Héros, ils procurent leur propre ruine, en mettant en relief leur impuissance pour extorquer les plans de Dieu: le diable va être berné.

L'euphémisation du mal: sa banalisation

C'est cette position actantielle qui prédétermine la banalisation du principe du mal; mais la ressource foundamentale pour aboutir à ce processus c'est la propre conception scénique des diables. Nous sommes devant la "diablerie".

La fonction de la "diablerie". La "diablerie" se caractérise par la présence du bruit, de la violence et parfois de l'obscénité. Ces éléments ont pour but de déterminer la différence de plans qui existe entre le bien et le mal: il n'est jamais question d'une relation d'opposition établie sur une identité de plans, mais il y a un clair décalage entre eux, fondé sur le manque de pouvoir de l'Espace infernal pour vaincre Dieu. Afin de remarquer cetter inferiorité les diableries se fondent sur des ressources qui pourraient être considérées propres à la Farce:

1. Les coups et les bagarres entre diables:

Sathan est battu par les diables quand il échoue dans ses efforts de tenter Jésus – vv. 17248–17287.

2. La présentation de Lucifer comme un personnage "trompé":

Puisque les diables ne sont pas parvenus á reconnaître la nature rédemptrice du Christ, ils ne se rendent pas compte que leurs démarches pour le perdre, entraînent leur propre destruction. Quand ils en sont conscients, Lucifer s'adonne á un désespoir comique et impuissant – vv. 28648–28695 – qui fournit le point de contraste au triomphalisme naïf de Sathan qui croyait avoir résussi à éliminer l'ennemi – 25581–25586 –.

3. La caractérisation de ces personnages se complète à travers leur langage: ils parlent plutôt comme des personnages sortis d'une taverne que comme des êtres qui incarnent un principe universel.

A notre avis, les diableries provoquent trois effets fondamentaux:

- Elles essaient de présenter des êtres agressifs, et en même temps ridicules, qui tombent dans le désespoir à cause de leur impuissance, conséquence d'une incapacité pour comprendre la réalite; cela les fait apparaître comme de pauvres naïfs qui ont travaillé à obtenir leur propre ruine.
- D'aileurs, ces efforts manqués ne fournissent qu'un plus grand point de contraste avec la victoire du Héros.
- Un dernier but, d'aprés Payen,[12] est d'offrir un prétexte pour le rire, à travers les scènes de coups qui président ces situations. Ce traitement du démoniaque relève d'un pittoresque monstrueux qui essaie de prêter plus au sourire qu'à la crainte.

Simone Sofian[13] considère ces "slapstick" et cette cruauté comique comme des ressources qui visent à amuser le public, et qui en même temps cherchent la banalisation de la figure démoniaque afin de permettre de soulager la peur que celle-ci provoquait.

La Fonction de mal dans le contexte d'un rite de transition

Nous voyons, donc, l'interrelation entre des épisodes drôles et une action qui institue une nouvelle cosmogonie. Le Mystère, étant donné son caractère de dramatisation de la fondation du monde moderne, pourrait être inscrit dans le contexte des rites de transition. Il n'a pas quitté son caractère de commémoration rituelle du changement de loi produit "in illo empore"; et s'il est joué, c'est pour renforcer cette "Nouvelle Alliance" entre Dieu et la Ville.

En tenant compte des études de V. Turner,[14] nous pourrions identifier le personnage du diable au "trickster"[15] ou personnage bouffon qui marque la transition d'un état à un autre dans le contexte de ces rites. À notre avis, ces éléments trémendistes provoquent la rupture avec une époque antérieure en même temps qu'ils focalisent la transition entre le monde chaotique et le nouvel ordre qui va se produire. En employant des mots de Reiner Hess, nous pourrions dire qu'ils sont des "marques mythiques qui focalisent l'action de la régénération".[16]

Ces défaites progressives de l'élément terrifiant provoqueraient sans doute quelque soulagement dans un public qui avait l'occasion de matérialiser la cause de leur peur à travers un système de représentation où le grotesque l'emporte sur le sérieux, afin d'éliminer justement ce côté réel qui n'obsédait que trop l'esprit de l'époque. Mais cette catharsis collective serait définitivement asséné aux figures du mal: leur destruction physique et l'abolition de leur pouvoir.

La destruction du mal

Il y a une double nature dans le héros christologique: dans la première partie du Mystère il est la victime propitiatoire qui se soumet au pouvoir temporel du mal – pouvoir des Juifs, animés initialement par les Enfers; mais cette mort, tel que nous l'avions indiqué auparavant, apparaît comme le pas préalable pour qu'il se constitue en héros épique qui descend aux Enfers pour vaincre les Diables.

Nous pourrions dire que le Diable a été l'élément indispensable du drame eschatologique et de la victoire sur la mort; mais c'est sa nature négative qui domine dans ce drame; de façon que la victoire définitive du héros s'exprime par une victorire physique sur lui.

L'Addition VII du *Mistere de la Passion* de Jehan Michel empruntée à l'évangile de Nicodème[17] présente cette action en recréant le topique du "tollite portas".

Les derniers efforts de Lucifer pour se défendre et empêcher l'entrée du Christ sont banals par rapport à l'action libératrice du héros qui apparaît à présent comme le guerrier qui en abbattant les remparts du mal, tranche les liens emprisonnants de celui-ci.

Sathan: "Je mectray roches de metal . . .
aussi grands que grands chasteaux" (vv. 28660–28661).
Les Dyables: "Haro, haro, haro, helas,
voicy ung terrible charroy!" (AD.VII, vv. 8–9).

Cette volonté de mettre an scène la destruction du mal pourrait être complétée par un autre Mystère, dans lequel le Christ anéantit l'autre entité maléfique, qui apparaît sous une caractérisation proche de celle des diables: Le Juif. Il s'agit du *Mistere de la Vengeance de Nostre Seigneur*.[18] Dans cette pièce le Chirst, à travers le peuple qui incarne par antonomase les Gentils – Rome – dètruit le Peuple d'Israël et le condamne au banissement.

Même si nous avions parlé de la nature bénéfique de la mort du Christ, les auteurs matériels de celle-ci deviendront des sujets maudits qui assument en exclusive la culpabilité de cette action: constituès en bouc-èmissaires ils seront détruits, ce qui assûre la tranquillité de la communauté qui ne devra assumer aucune responsabilité dans la mort de Dieu, même s'il est mort pour la racheter du péché.

Le mal a été maîtrisé sous ses différentes manifestations; il a été vaincu ou, du moins, les armes pour le faire – pouvoir spirituel et temporel, qui sont les structures sur lesquelles se fonde cette communuaté – ont été fournies. Les métaphores qui incarnent les puissances maléfiques dans le Mystère – Lucifer et sa cour, les Juifs –, tissent le réseau d'opposition affronté par le héros – le Christ –, qui tient non seulement le rôle de Messie Universel, mais celui de héros emblématique de la cité; c'est ainsi que le récit biblique, sans perdre totalement sa signification, se transmue en histoire d'une réalité sociale, qui s'exprime à travers ses unités culturelles fondamentales.

Le processus allégorique se fonde sur une abolition temporelle entre l'époque du mythe fondateur et le temps contemporain, si bien qu'il y a une assimilation entre le rècit biblique et l'histoire de la ville, qui permet l'intégration des conflits de l'époque dans l'historie christologique, en rendant possible l'appropriation du triomphe du héros sur le mal de la part de la communuauté. L'action du héros vainqueur dans le Mystère, constituée celle-ci en nouveau rècit exemplaire, devient la garantie d'une victoire éternelle de la communauté urbaine sur ses avatars.

Le mal a été invoqué sur la scène pour pouvoire le chasser.

Université Autonome de Madrid

NOTES

[1] Eco, U. *Lector in fabula. La coopération interprétative dans les textes narratifs* (Paris: Bernard Grasset, 1985).
[2] Groupe, M. *Rhétorique générale* (Paris: Le Seuil, 1982).
[3] Eco, U. *La estructura ausente* (Barcelona: Lumen, 1986).
[4] Payen, J. -Ch. "Théâtre médiéval et culture urbaine" in *Revue d'histoire du théâtre*, **35** (2) (1983): P. 233–250.
[5] Durkheim, E. *Les formes élémentaires de la vie reliqueuse* (Le système totémique en Australie) (Paris: Alcan, 1912).
[6] Sorokin, P. *Dinámica social y cultural* (Madrid: Instituto de Estudios Politicos, 1962).
[7] Muchembled, R. *Culture populaire et culture des élites dans la France Moderne (XVI-XVIII)* (Paris; Flammarion, 1978), p. 198.
[8] Warren, G. "Diables Passion Gréban" in *Chimères*, **15** (1) (1981): p. 23–35.
[9] Gréban, A. *Le Mistere de la Passion*, éd. O. Jodogne (Académie royale de Belgique, 1965).
[10] Turner, V. *From Ritual to Theature: The Human Seriousness of Play* (New York: Performingarts, 1982).
[11] Michel, J. *Le Mistere de la Passion de Nostre Seigneur*, éd. O. Johogne (Genève; Droz, 1979).
[12] Payen, J.–Ch. "Pour en finir avec le Diable médiéval ou pourquoi poétes et théologiens du Moyen Âge ont-ils scrupule à croire au démon?" in *Sénéfiance*, **6** (1979), p. 401–425.
[13] Sofian, Simone Lotuen. *The Role of the Comic in the Mystère and the Miracle* (University of Cincinnati, 1985).
[14] Turner, V *La selva de los mitos* (Madrid: Siglo XXI, 1980).
[15] Turner, V. *op. cit.*
[16] Hess, R. *El drama religioso románico como comedia religiosa y profana* (Madrid; Gredos, 1976).
[17] Migne, P. *Dictionnaire des Apocryphes*, rprl. Brépols, Turnhout, 2 vol. (Belgium, 1989).
[18] Mercadé, E., *Vengance de Nostre Seigneur JesuChrist* (Paris: Alain Lotrian, 1539).
[19] *Le Mystere de la Saincte Hostie* (Air: A. Pontier, 1817).

II. L'ALLÉGORIE DU MAL DANS *LA FIN DE SATAN* DE VICTOR HUGO

Margarita Alfaro Amieiro

Le Fin de Satan,[1] vaste poème annoncé dans la Préface de *La Légende des Siècles*, commencé en 1854 et achevé en 1860, s'intègre dans le cadre idéologique du romantisme[2] où la vision de la philosophie de Hugo et de sa production littéraire met en relief:
- le dualisme à tension tragique implicite dans l'homme,
- et son refus vers n'importe quelle oppression dogmatique.

Tout le discours hugolien se construit à l'intérieur d'une progression dynamique.

Initialement étaient prévus trois livres clairement différenciés bien que le troisième probablement intitulé *La Prison* n'ait jamais été écrit.[3]

Les parties présentes dans le poème actuel décrivent un processus d'allégorisation en rapport avec la figure de Satan: la création et la défaite progressive du mal sur la terre et sa projection positive pour atteindre le bien et la liberté.

"Hors de la Terre I", "Le Glaive", "Hors de la Terre II,", "Le Gibet", "Hors de la Terre III" et "L'Ange Liberté" s'articulent en trois tableaux où se met an évidence la construction des "structures dramatiques"[4] du personnage impliqué. Certes, Hugo récupère dans ce poème la tradition biblique judéo-chrétienne, et d'apprès les mots de Ch. Baudouin "le romantisme est héritier de toute la dramatisation de la littérature biblique, de l'iconographie médiévale et du Paradis Perdu de Milton".[5]

Dans notre étude nous allone mettre en évidence la progression de ce fil conducteur à l'intérieur des mouvements suivants:

1. La genèse de Satan: cosmovision où a lieu un processus de création par opposition; c'est-à-dire l'acceptation de l'existence d'un état positif idéal face à un processus de déchéance, moteur du poème. En conséquence certaines situations antithétiques se problématisent et créent deux images contraires polarisées: lumière-ténèbres et ciel-abîme. D'autres manifestations fondamentales s'intègrent à l'intérieur de ce dualisme en tension: à savoir, le parcours de la chute, la transfiguration de Satan et sa pulsion ascensionnelle.
2. La projection de Satan: installation du mal sur la terre.[6]

3. Sa réactualisation de signe positif à partir de l'incident de la plume qui a échappé à la chute.

Une fois établie l'importance de ces trois mouvements nous passons à leur analyse au détail:

1. *La genèse de Satan*

L'allégorie de la création du mal peut être articulée à partir de certaines structures diamétrales:

1.1. *"Et nox facta est"* L'importance de l'origine de Satan est essentiellement fondée sur le principe de l'exclusion qui génère une dyade lumière-ciel et son opposition antithétique obscurité-gouffre. La force génératrice du poème est l'inversion de "Et nox facta est" face à "Et lux facta est", premier moment de la création divine. À ce propos Georges Gusdorf dans *L'homme romantique* (1984) considère que "le romantisme fait élection de domicile dans le domaine de la nuit; il oppose aux prestiges aveuglants des lumières la clarté vraie de l'illumination: la mort devient une situation limite dans le règne des ténèbres . . . La fascination de la nuit tient à ce qu'elle propose le cheminement naturel pour aller de l'*évidence* du dehors à l'*invidence* du dedans, exigence supérieure à toutes les sollicitations de l'extériorité".

Dans ce poème l'image des ténèbres est conçue comme une disparition progressive de la lumière, emblème initial de Lucifer, à conséquence d'une malédiction fatale: "maudit, autour de toi les astres s'éteindront". (p. 40)

En tout cas il s'agit d'un renversement cher à Hugo puisque nous trouvons le même parcours, de Nox à Lux, dans *Châtiments*.

Dans le troisième mouvement la force envahissante de la lumière, dont le porteur est l'Ange Libertè, fera possible une double renversion: ténèbres-lumières et mal-bien, ou disons, suivant les implications du processus d'allégorisation considéré, tyrannie-liberté.

1.2. *Le parcours de la chute.* La durée et la progression de la chute apparaissent dès le premier moment comme une situation obsédante renforcée par tous les éléments archétypaux qui configurent l'image du démon, un être banni, refusé et damné. La punition est générée par un ordre impératif qui marque l'exclusion: "Quelqu'un, d'en haut, lui cria:

'Tombe!'" (p. 38). Ce verbe déclenche un mouvement de descente mis en relief par la présence et la répétition de sèmes appartenant au champ sémantique de la chute: "tomber", "s'enfoncer", "tomber plus bas".

À côté nous observons la présence répétitive de l'imparfait qui renforce l'aspect duratif du parcours, action lente, progressive et inévitable.

La déchéance irréparable est doublement réaffirmée, d'un côté par les sentiments éprouvés d'intensité dramatique puisqu'il s'est produit une renversion douloureuse par rapport à l'état antérieur, idéal et positif, et de l'autre côté par des images évocatrices qui soulignent l'horreur de l'éloignement et la gravité de la chute: "Il s'enfonçait dans l'ombre et la brume", "Il tombait foudroyé, morne, silencieux,/ Triste, la bouche ouverte et les pieds vers les cieux,/ L'horreur du gouffre empreinte à sa face livide" (p. 37).

Dans ce parcours, la descente apparaît comme tragique et fatale, il y a une volonté pour l'éviter et pour trouver un point d'arrêt d'où regarder vers le haut et contempler le soleil et la lumière:

> "Il tombait. Tout à coup un roc heurta sa main;
> Il l'étreignit, ainsi qu'un mort étreint sa tombe,
> Et s'arrêta"
> (p. 37).

Dans la description de l'arrêt se cristallise la renversion négative de l'acte de création divin. Satan lance un cri fatal, celui de "Mort!", qui déchaîne l'acte créateur du mal et la personnification de principes abstraits négatifs: Caïn, la mort, Judas, le mensonge et Barabbas l'esprit satanique.

Après cet arrêt, à conséquences maudites, la chute recommence et le processus d'éloignement par rapport à la lumière et au ciel apparaît très distinctement présidé par l'horreur de la déchéance sans issue. L'entourage possibilite cette chute vers le néant:

> "La chute du damné recommença. – Terrible,
> Sombre, et piqué de trous lumineux comme un crible,
> Le ciel plein de soleils s'éloignait, la clarté
> Tremblait, et dans la nuit le grand précipité,
> Nu, sinistre, et tiré par le poids de son crime,
> (. . .)

Tout à présent
Le fuyait; pas d'obstacle à saisir en passant,
Pas un mont, pas un roc croulant, pas une pierre,
Rien, l'ombre, et d'épouvante il ferma sa paupière".
(p. 38)

Le parcours de la chute a une fin dans l'espace, les sommets de l'abîme, "les cimes du chaos". Satan est le premier et le seul habitant dans ce paysage de désolation et de ruine, c'est lui qui fend les ténèbres: "(Il) Tombait, et, comme un coin, sa tête ouvrait l'abîme/ Plus bas! plus bas! toujours plus bas!" (p. 38). Son voyage a permis d'établir la topographie des ténèbres jusqu'à maintenant inexistante par rapport au ciel et à la lumière. La Création du mal, conséquence de la chute, est accomplie.

1.3. *La transmutation de Satan.* La transformation négative de Satan est dominée par la métamorphose de son état initial, ange porteur de lumière, et par la présentation d'un autre état où les traits de la monstruosité physique sont fonctionnels, il y a une animalisation résultante: "la chauve-souris du cachot éternel":

"Tout à coup il se vit pousser d'horribles ailes;
Il se vit devenir monstre, et que l'ange en lui
Mourait, et le rebelle en sentit quelque ennui".
(p. 39)

Dans cette description le front, repris plusieurs fois tout au long du poème, est le porteur des marques distinctives de Satan – pour l'ange Liberté le front sera le porteur de l'étoile lumineuse capable de dissoudre l'esprit du mal.

Les ailes[7] finalement suivent un renversement emblématique. Dans la condition antérieure, élément aérien et esthétique positif, dans la nouvelle situation elles deviennent une partie monstrueuse et sauvage: "horribles ailes", "aile affreuse", "aile onglée et chauve", capables de voler, mais inutiles pour surmonter le gouffre.

L'absence de plumes marque la différence entre la positivisation ou la négativisation de cet élément. Les plumes se perdent dès le premier moment de la chute:

"Seul et derrière lui, dans les nuits éternelles,
Tombaient plus lentement les plumes de ses ailes".
(p. 37)

Métaphoriquement elles se transforment en "des ailes-chaînes"[8], raison de son impuissance et de son nouvel statut de prisonnier et de damné infâme dans l'abîme.

1.4. *Le gouffre et les ténèbres: la pulsion ascensionnelle.* L'opposition ciel-grouffre se présente comme une progression vers une "opaque obscurité", "une ombre universelle" et finalement "les nuits èternelles" où le conflit de force et de domaine est évident: "Il aura le ciel bleu" vs. "moi, j'aurai le ciel moir" (p. 39).

Le seul point de connexion possible avec son état initial est l'oeil fixé[9] vers "les astres"; les "ténèbres visibles", expression miltonienne, sont un soulagement à sa situation dramatique.

L'éloignement physique comporte un affaiblissement progressif et implacable de la lumière. La description de cet éloignement est une graduation où l'enchevêtrement avec l'élément antérieur renforce la progression:

"Le ciel plein de soleils s'éloignait, la clarté
Tremblait, . . . "
"Quand il rouvrit les yeux, trois soleils seulement
Brillaient, et l'ombre avait rongé le firmament.
Tous les autres soleils étaient morts".
"Les trois soleils, de loin, ainsi que trois prunelles
Le regardaient, et lui ne les regardait pas".
"Les trois soleils mêlaient leurs trois rayonnements".
"Soudain il tressaillit; il n'en restait plus qu'un"
"Et déjà le soleil n'était plus qu'une étoile"
"Et tout disparaissait par degrés sous un voile
(. . .)
L'étoile maintenant n'était qu'une étincelle"
"L'étincelle
N'était qu'un point rougeâtre au fond du grouffre obscur"
"L'astre au fond du brouillard, sans vent qui le ranime se refroidissait, morne et lentement détruit"

"L'astre était presque noir"
"Et cria: – Désespoir! le voilà qui pâlit!
" . . . et l'astre s'éteignit".
(pp. 43–44)

Le point d'union entre Satan et le monde s'affaiblit progressivement jusqu'à la disparition totale de la lumière comme une mort angoissante, une agonie lente qui lutte pour s'empêcher de mourir:

> "Et l'astre agonisait sous ses regards farouches
> Il mourait, il luttait. Avec ses sombres bouches
> Dans l'obscurité froide il lançait par moments
> Des flots ardents, des blocs rougis, des monts fumants,
> (. . .)
> Comme si ce volcan de vie et de lumière,
> (. . .)
> N'eût pas voulu mourir sans insulter la nuit
> Et sans cracher sa lave à la face de l'ombre".
> Dans L'obscurité froide".
> (pp. 43–44).

Cette mort sépare définitivement Satan du monde inondé par la lumière. L'absence de lumière donne naissance à l'ombre de l'abîme qui nous renvoie au Mythe de la Caverne platonicien où l'image gnostique des ténèbres et de la lumière instaure la bipartition entre le Bien et le Mal.[10]

Nous devons signaler, en plus, que dans le texte biblique l'abîme équivaut à un océan, à une mer originaire[11] et dans le poème de Hugo eaux et ténèbres participent de cette image isoomorphe: "Et l'archange comprit, pareil au mât qui sombre,/ Qu'il était le noyé du déluge de l'ombre". (p. 44) Cela explique l'image du "nageur" qui désespérement essaie de sortir des eaux sans succès: "Satan, comme un nageur fait un effort suprême,/ Tendit son aile onglée et chauve, et, spectre blême,/ Haletant, brisé, las, et, de sueur fumant,/ Il s'abattit au bord de l'âpre escarpement" (p. 43).

De la part de Satan nous remarquons l'existence d'une pulsion ascensionnelle plaintive et suppliante, inverse au processus de la chute. Les procédés verbaux d'énumération marquent l'impuissance de Satan:

"Il courait, il volait, il criait: – Astre d'or!
Frère! attends-moi! j'accours! ne t'éteins pas encor?
Ne me laisse pas seul! –".
(p. 41)

L'idée d'élévation et même de conservation de l'élément lumière se constitue en moteur d'un effort bouleversant de la part de Satan. "La lumière étincelante" le pousse à lutter pour obtenir son objectif:

"L'astre parfois semblait s'éteindre et s'éclipser,
Et l'horreur du tombeau faisait frissonner l'ange;
Puis une clarté pâle, obscure, vague, étrange,
Reparaissait, et l'ange alors disait: Allons".
(p. 42).

La fatigue physique résout en impuissance l'effort démesuré et angoissant de son vol inscrit dans le temps: "Il vola dix mille ans, Pendant dix mille années".[12] Il semble rester irrémédiablement liè á une nature négative, expression de la fatalité qui coïncide avec le mal dans toute la pensée hugolienne.

Toutes les images des conditions externes, – la rudesse de l'hiver –, et la représentation sensorielle – sensations auditives, cromatiques et visuelles – renforcent le fatalisme angoissant de la genèse du mal.

2. *Projection de Satan: installation du mal sur la terre*

Satan se constitue en moteur négatif de la création divine, sa conduite est renversée par rapport à celle du Créateur: "Pensez-vous quelquefois à ce que fait l'archange,/ L'Être d'en bas? Il est le Méchant. Il s'en venge./ Il prend l'âme, la vie et le jour à revers;/ Et de sa chute il fait celle de l'univers" (p. 115).

Isis-Lilith,[13] sa fille, est la figure allégorique qui incarne cette projection de l'esprit du mal sur la tarre. Dieu déchaîne le déluge avec une fonctionnalité purifiante sur la souillure, cette fonction se voit invalidée par la force maléfique et destructive de Lilith qui brandit les trois éléments symboliques du mal qui n'ont pas été étruits par le déluge; le fer, le bois et la pierre.

Ces trois images donnent lieu à des épisodes où est présente la description de la déchéance du monde. Dans la pensée hugolienne, la dégradation du bonheur et du progrès de l'humanité.

À l'intérieur de ce conflit Satan est le germe de deux formes de tyrannie: la première, la domination par la force physique génère une oppression politique depuis l'histoire de Rome. Nemrod,[14] personnage biblique, anime l'esprit de destruction, raison du malheur de l'humanitè, et illustre l'allègorie de l'orgueil. Sa révolte contre Dieu se tourne en auto-destruction, la flèche lancée vers le haut en est clairement l'élément symbolique.[15]

La deuxième met en question la tyrannie spirituelle exercée par la religion: "Car leur religion donne à l'humanité/ Une difformité misérable et terrible" (p. 124).

La dualité Satan-le temple vs. le Christ est le point fondamental qui marque l'écart entre la conception des Mystères et celle du poème de Victor Hugo. Dans *La Fin de Satan* l'église apparaît comme un opposant au héros messianique, le temple n'exerce pas une domination sur Satan, bien au contraire il est son allié, il est I'élément tyrannique qui détruit Dieu. La mort du Christ n'a pas les valeurs rédemptrices qui sont célébrées aux Mystères.

Dans le poème objet de notre étude, Satan domine par une élection de l'humanité infâme. Les valeurs antithétiques antre Satan et Dieu sont projetées vers une positivisation implicite de Satan. Sa condition première offrira au monde la libération, c'est lui qui deviendra par la suite le racheteur de la souillure des hommes.

Nous citons les derniers vers du monologue de Barabbas épouvanté devant Jésus cruxifié, les ténèbres dominent en ce moment la terre:

> "Peuple, affreux peuple sanglant,
> (. . .)
> "Malheur, monde impur, lâche et rude!
> Monde Où je n'ai de bon que mon ingratitude,
> Puisse à jamais ce Christ sur ta tête saigner!
> Qu'un déluge d'opprobe et de deuil t'engloutisse,
> Homme, plus prompt à choir du haut de la justice
> Que l'éclair à tomber du haut du firmament!
> Sois maudit dans ces clous, dans ce gibet fumant,
> Dans ce fiel! sois maudit dans ma chaîne brisée!
> Sois damné, monde à qui le sang sert de rosée,
> Pour m'avoir dèlivré, pour l'avoir rejeté,

Monde affreux qui fais grâce avec férocité,
Toi dont l'aveuglement crucifie et lapide,
Toi qui n'hésites pas sur l'abîme, et, stupide,
N'as pas même senti frissonner un cheveu
Dans ce choix formidable entre Satan et Dieu!".
(p. 176)

3. *La réactualisation de Satan*

La réhabilitation de Satan commence avec l'épisode de la plume qui a échappé à la chute. La plume devient la médiatrice de sa récupération pour un nouvel statut: celui de l'ange liberté.

Tout au long du poème nous avons pu observer la permanence d'une pulsion ascensionnelle significative et pertinente. La première était celle de Satan vers la lumière impossible et la deuxième celle de Nemrod vers le ciel, l'équivalent humain de la chute de Satan. La descente vers le gouffre de l'Ange Liberté sera justement le renversement souhaitè pour que la pulsion ascensionnelle se matérialise comme un acte libérateur positif.

Dans le cadre de cet élan ascensionnel s'intègrent des manifestations antithétiques – nuit-vie, jadis-maintenant, angoisse-acceptation doubloureuse – qui déterminent chez Satan la dualité polarisée entre le bien et le mal. La lutte intérieure se manifeste momentanément par des réactions d'autoréaffirmation et d'orgueil. Néanmoins l'essence positive de Lucifer n'a pas été annihilée:

"Jadis, ce jour levant, cette lueur candide,
C'était moi. – Moi! – J'étais l'archange au front splendide,
(. . .); maintenant
Je suis l'astre hideux qui blanchit l'ossuaire.
Je portais le flambeau, je traîne le suaire;
(. . .)
Tout devant moi, vers qui jadis l'amour vola,
Recule et fuit.
(. . .)
Oh! je l'aime! c'est là l'horreur, c'est là le feu!
Que vais-je devenir, abîmes? J'aime Dieu!
Je suis damné!"
(p. 188)

L'image du mal va suivre dans ce mouvement un trajet progressif justement en contraste avec la naissance du mal et de sa nature destructive.

Toute la progression du nouveau personnage s'articule autour de deux dominantes ambivalentes qui seront définitivement polarisées positivement dans la projection de l'ange liberté:

La première: le pouvoir de la haine vs. un vouloir amour pour atteindre le bien.

Il y a une volonté dominante pour avilir la haine. La capacité de pensée et dans cette capacité la reconnaissance de la haine comme réalité qui nous enchaîne sera pour Hugo la force génératrice de la délivrance: "Haïr délivre" (p. 215).

La Deuxième: ténèbres vs. vouloir atteindre la lumière.

L'absence de l'élément lumineux est déterminante, elle symbolise le néant absolu, la renversion se fera au moyen de la présence de la lumière.

Ces deux dominantes sont antithétiques dans leur structure et provoquent une polarité dans le discourse de Satan; il y a une alternance entre un vouloir – faire et un non-pouvoir-faire qui détermine une situation dramatique d'angoisse et de souffrance: "Ne pouvoir remonter, même quand je le veux!/ Je suis le misérable à perpétuité". "ô ma lumière, où donc es-tu? Satan t'implore/ (. . .) reviens, aurore, aurore, aurore!/ (. . .) ne me dis pas: jamais!/ Je souffre! – Oh! tout est noir, je ne vois pas, je hais!" (p. 200).

Hugo laisse entrevoir une vision manichéiste du monde où la bipartition n'est pas souhaitable, l'équilibre entre le mal et le bien doit être transformé en faveur de la liberté:

"Grâce! pardonne-moi! rappelle-moi! prends-moi!
Grâce! Ne sens-tu pas qu'il faut que toute chaîne
Se rompe, et que le mal finisse, et que la haine
S'éteigne, évanouie en ta sérénité?
Quoi! le bien infini, le mal illimité!
Toi le bien, moi le mal! est-ce que c'est possible?
Y songez-vous, Seigneur? un partage entre nous!"
(p. 204)

L'Ange Liberté est présenté paradoxalement comme l'identité de Satan: "Ce qui survit de toi, c'est moi. Je suis ta fille" (p. 236). Leur identité et leur altérité permettra la coexistence en harmonie.

La figure de l'ange et de Lilith[16] incarnent une gémélité antithétique qui triomphe au profit du principe du bien. Lilith, élément corrupteur du monde, détient la féminité mythique des déesses terribles et sanguinaires et s'oppose à l'ange vierge, métaphore de la pureté et symbole de la transmutation du mal en bien. Baudouin[17] établit une série de catégories d'images isomorphes, présentes dans le poème, telles que: jour, clarté, azur, rayon, vision, grandeur et pureté relevantes dans la figure de la femme rédemptrice; celle-ci est clairement mythifiée par le romantisme français.

Cette fonction rédemptrice se verra complétée par l'acceptation de Satan et réclamée par l'ange liberté au nom de la liberté et de la justice.

L'espace abissal est à mettre en rapport avec la claustration où l'absence de sommeil et de rêve est conçue comme une damnation et même comme une mort – positive par la suite au sens donné par Novalis comme une initiation ou comme un chemin à commencer. Le sommeil possibilite justement la projection de Satan, il devient l'espace qui permettra la libération de l'être: "le rêve est la voie d'accès, un état intermédiaire entre le sommeil et la veille"[18]. L'impératif "Va!", antithèse de "Tombe!" engendre ce moment bienfaiteur de la lutte cosmique et de l'épopée satanique: "Le facteur commun aux *Contemplations*, à *La Légende des siècles* et à *La fin de Satan* se résumerait en cette équation: Hugo sauvé – l'Homme sauve – Satan sauvé. . ."[19]

Pour conclure disons que Victor Hugo nous a présenté dans le cadre de la téophanie biblique la personnification et la représentation de différentes images allégoriques construites sur la notion de cohésion antagoniste et antithétique, ce qui explique sa vision de l'histoire comme dynamicité, et où s'intègrent les réflexions sur le mal, la cosmogonie du pardon et la genése idéale d'une humanité libre en dehors des structures établies. De là l'explication d'une révolution où l'expérience avec Dieu et avec Satan, celui-cihéros messianique, soit le fondement d'un monde en équilibre où l'on admett le progrès.

NOTES

1 Nous suivons l'édition: Victor Hugo *La Fin de Satan* (Paris: N.R.F. Gallimard 1984).

2 Pour saisir la structure idéologique sous-jacente dans la création hugolienne nous renvoyons à l'étude de J. del Prado: *La Préface de Cromwell: un texto reaccionario*" (U.C.M., 1885).

[3] Victor Hugo écrit en 1854 ce poème qui est le résultat d'une élaboration progressive de différents fragments modifiés tout au long de six annéns de travail. Sur l'établissement du manuscrit définitif et les sources d'inspiration *Cfr.* l'étude faite par Max Milner, *"Victor Hugo et la Fin de Satan"* in *Le Diable dans la littérature française de Cazotte à Baudelaire, 1772–1861*, T. II (Paris: Librairie José Corti, 1960), pp. 359 et ss.
[4] Expression citée par G. Durand. *Cfr.* son étude sur *"Dualité et drame des Winneabago à Victor Hugo" in* L'âme tigrée (Ed. Denoe) (1980) pp. 75–116.
[5] C. Baudouin, *Le Triomphe du héros* (Paris: Plon, 1952), p. 143.
[6] Victor Hugo en écrivant *le Fin de Satan* fait des reflexions personnelles sur l'origine et les conséquences du mal.
[7] La plume, élément recurrent hugolien, apparaît dans des poèmes divers entre 1854–1855: ""Océan" (*La Légende des siècles*, XLI-1854 –; "Claire" (*Les Contemplations*, VI, 8 – 1854 –; "À celle qui est voilée' (*Ibid.* VI, 1855; Les Mages" (*ibid.* VI, 23 – 1855).
[8] Ici le symbolisme des ailes est évident. Dans la mythologie occidentale elles représentent une force spirituelle euphémisée. Les ailes des animaux nocturnes renversent la qualité de cette force positive. Les ailes artificielles d'Icarus donnent raison de son incapacité pour remplir la fonction souhaitée, ce qui explique notre interprétation "d'aile-chaîne". *Cfr.* le mythe d'Icarus et de Prométhée.
[9] Le motif de la contemplation est recurrent tout au long du poème.
[10] D'après G. Durand, *op. cit.*, p. 97, le mythe de la caverne platonicien est ramassé par l'antithèse hugolienne lumière-ombre et assimilée au bien-mal.
[11] *Cfr.* Gn 1, 2.
[12] Dans le NT Satan a été banni dans l'abîme pour un royaume de mille ans. *Cfr*, Ap. 20 1.3.
[13] Gémélité antithétique, celle de Lilith-Isis. Lilith, esprit malfaiteur et Isis déesse égyptienne bénéfique. Cette assimilation, d'après Paul Vernière, s'était produite depuis le XVIII siècle en dépit d'Isis.
[14] Dans la mythologie biblique traditionnelle Nemrod est considéré comme le descendant de Noé. *Cfr.* G. x, 9 Dans le poème de Victor Hugo il est décrit sous le prisme de la monstruosité et du gigantisme. Ses actions sont marquées par la négativité. La tradition biblique explique que la tour de Babel faisait partie de l'empire de Nemrod G. X, 10. Nemrod est orgueilleux et se révolte contre Dieu, celui-ci exerce sa vengeance et détruit la tour de Nemrod, de là la confusion des langues dans toute la terre. (L'étymologie de Nimrod explique cette acception du terme.)
[15] Certains exégètes considèrent les flèches souillées de sang par les anges pour faire croire qu'elles avaient blessé Dieu.
[16] *Cfr.* l'étude de L. Cellier, *L'épopée romantique* (Paris; P.U.F. 1954), pp. 234–245.
[17] *Cfr.* C. Baudouin, *Psychanalyse de Victor Hugo* (Edit. Mont-Blanc, 1944), p. 202.
[18] *Cfr.* G. Gusdorf, *L'homme romantique* Paris: Payot, pp. 181–219.
[19] L. Cellier, "Chaos vaincu", *Victor Hugo et le roman initiatique, Parcous initiatiques* (Neuchâtel; La Baconnière, 1977), p. 164.

CONCLUSION

Nous avons étudié deux processus allégoriques fondés sur la même matière thématique et pourtant les valeurs que celles-ci accordent au topique littéraire de Satan sont contradictoires. Dans le premier cas Lucifer et sa cour sont vaincus par le Christ. C'est ainsi que le principe du mal est affaibli. Mais dans l'oeuvre hugolienne l'euphémisation de l'image du démon est beaucoup plus forte: il est nié; il disparaît pour devenir une entité isomorphe de Dieu.

L'explication de ces différentes dynamiques peut être accordée en fonction de la nature des processus allégoriques: les deux allégories se fondent sur le même récit cosmologique – la matière biblique –, mais la grande différence c'est que chaque groupe – les Mystères et le poème de Hugo – est devenu à son tour une cosmogonie particulière qui explique l'univers de ses auteurs et de ses époques: le premier gardant et particularisant le sens littéral; le deuxième en opérant une subversion.

Les Mystères, tel que nous l'avons dit, sont l'expression d'une société affectèe par le fantôme de l'instabilité et obsédée pour défendre ses structures hiérarchiques et spirituelles, en détruisant tout ce qui puisse les ménacer: le péché, "l'autre". Le diable incarne ces valeurs; et la volonté de le détruire répond à une conception duelle de l'existence qui refuse la synthése des principles contraires.

Le Romantisme, par contre, fait un effort syncrétique pour établir un équilibre entre le bien et le mal; dans cette époque "la Révolution française, règne de Satan, cesse d'être l'ère de la désolation, pour devenir l'heure sainte:[1] la révolution est valorisée positiviment, et avec elle, Satan, le grand rebelle.

Cela justifie un déplacement significatif qui présente le mal incarné plutôt par les structures hiérarchiques qui enchaînent l'homme, – et que le Moyen Âge s'acharnait à défendre – que par le personnage de Satan, qui apparaîtra finalement en élément libérateur.

La perspective tragique qui caractérise Satan dans cette oeuvre s'écarte de la volonté banalisatrice des Mystères; et la fonction dégradatrice du diable dans la première partie de *La fin de Satan* n'est que le point d'antithèse qui met en relief le processus antiphrasique qui va s'opérer dans le texte. La négativité de Satan est niée et avec elle la propre figure du démon. Satan devient le père de l'Ange Liberté et l'émetteur d'un agent qui affranchira les hommes du mal – les ordres caducs, les ordres contraignants –, c'est le créateur d'un nouvel ordre.

Nous nous sommes trouvés devant deux personnages différents, parce que le contexte littéraire dans lequel ils se développent présente deux cosmogonies différentes; et parce que les circonstances qui donnent naissance à l'un et à l'autre répondent à des besoins vitaux bien opposés.

Université Autonome de Madrid

NOTE

[1] Cellier, L., *L'épopée Romantique* (Paris: P.U.F., 1954), p. 57.

HELEN FLETCHER THOMPSON

FROM METAPHOR TO MATTER: SARTRE, MERLEAU-PONTY AND THE SUBJECT OF SIGHT

At the start of *Being and Nothingness*, Jean-Paul Sartre inverts the progression of the cosmogony enclosed by those two terms: " . . . we must be careful never to posit nothingness as an original abyss from which being arose."[1] Instead, nothingness is imposed upon a being which can be situated "before" because it is profoundly atemporal, or, rather, indifferent. Sartre states: " . . . we could say that every negation is determination. This means that being is prior to nothingness and establishes the ground for it . . . This is what we mean when we say that *nothingness haunts being*."[2] Here, Sartre's nothingness is no more than a "haunting" which claims less density than light or shadow, assuming rather the flickering insubstantiality of the *event* motivating this ontological prioritization and the concurrent determinations which make a world.

Sartre amplifies the effect of nothingness upon being in his discussion of consciousness, where he states:

> The in-itself is full of itself, and no more total plenitude can be imagined, no more perfect equivalence of content to container. There is not the slightest emptiness in being, not the tiniest crack through which nothingness might slip in. The distinguishing characteristic of consciousness, on the other hand, is that it is a decompression of being.[3]

When the evanescent ground of the ontological event of prioritization assumes the *density* of self-identical being, the relation between in-itself and for-itself is materialized as one of impossible *passage*. The perfect self-identity of "content to container" will always deny a passage where nothingness might "slip in"; necessarily so, as the upsurge of the for-itself in the world is defined and sustained by this very impossibility. Thus the "decompression of being" marks not a *process* by which matter dissipates but the relation of consciousness to the density of that being, never a modification of the in-itself but an approach or a retreat from it. Sartre polices this impossible passage by refuting any chance of exchange between for-itself and in-itself, by definition: " . . . the For-itself adds *nothing* to the In-itself except the very fact that *there is* In-itself; that is, the affirmative negation."[4] The for-itself gives the in-itself nothing but its being, its "is": subsequently this being is both

M. Kronegger and A-T. Tymieniecka (eds.), Analecta Husserliana XLII, pp. 185–201.

inviolate and receptacle, eternally waiting for the project of "thises" projected upon it and eternally unmarked. Here Sartre's cosmogony recalls that of Plato's *Timeaus*, where Plato defines the "universal nature which receives all bodies," reminiscent of the in-itself because it is characterizable only by the perfect plenitude of self-identity:

> . . . that must be always called the same; for, while receiving all things, she never departs at all from her own nature and never, in any way or at any time, assumes a form like that of any of the things which enter into her; she is the natural recipient of all impressions . . . the matter in which the model is fashioned will not be duly prepared unless it is formless and free from the impress of any of those shapes which it is hereafter to receive from without.[5]

Both Plato's receptacle and Sartre's in-itself receive but cannot *retain*; total self-equivalence becomes equated with an ideal receptivity never compromised by the impressions imposed upon it from "without".

However, while Plato cannot conceive of any juncture where this "universal nature" might threaten the subject with an inarticulable transition, or dissolution, this project of passage is explored by Sartre in the face of the assertion that buttresses *Being and Nothingness*: "If it is impossible to pass from the notion of being-in-itself to that of being-for-itself and to reunite them in a common genus, this is because the *passage in fact* from the one to the other and their reuniting cannot be effected."[6] While the impenetrable face of the in-itself does claim an aloofness like that of the Platonic universal nature which never departs from herself, the Sartrian for-itself is incessantly reminded of the project of passage – through its own body as revealed in the gaze of the Other, in the corporeal density of its own past, and ultimately in the desires to have and to be. Sartre is ultimately confronted with this passage as the final insidious threat to being, a threat which erupts in the progression of the site of metaphor from the human body to pure matter.

In his early discussion of the project of passage, Sartre portrays the realization of physical desire as an erosion of the particularities of the body that would almost preclude gender:

> . . . breasts, buttocks, thighs, stomach: these form a sort of image of pure facticity. This is why also the true caress is the contact of two bodies in their most fleshly parts . . . the full pressing together of the flesh of two people against one another is the true goal of desire.[7]

In the end, Sartre will not figure the "masculine" and the "feminine" in an explicit physiognomy or act but rather in the process of their very dissolution, a process which in the case of sexual desire is arrested by the inevitable return of the Other to impenetrable objecthood. It is in Sartre's final metaphors of possession, in which human physiognomy is replaced by pure states of matter, that features of sexual difference can be inscribed upon the blank surface of the in-itself. *Quality*, and not physiognomy, becomes the site of attempted passage where Sartre re-inscribes *one* gender in a matter whose mute blindness becomes finally groping, fatally seductive; yet before reaching this end Sartre pursues controlled changes of state from solid to liquid, solid to gas.

Sartrian "smoothness" is at the hinge of physiognomy and quality, characterizing both the marble flank of the beloved and the pure quiddity of water. As Sartre states:

> What is smooth can be taken and felt but remains no less impenetrable, does not give way in the least beneath the appropriative caress – it is like water. This is the reason why erotic descriptions insist on the smooth whiteness of a woman's body. Smooth – it is what re-forms itself under the caress, as water re-forms itself in its passage over the stone which has pierced it.[8]

This smoothness again refutes the possibility of the (male) lover's dissolution, for *nothing* of his hand is added to the in-itself, whose smoothness testifies to its inability to retain any trace of the for-itself for itself. Blankness is perfect testimony, and water, incompressibly dense and infinitely yielding, appears as the ideal matter which can be possessed and yet which bears witness only to its own presence. Water is at the heart of Sartre's next phase of matter, a smoothness become vast and traversable and a metaphor gleaming with the innocence of an Alpine slope, that of snow. The matter of snow would *be* Sartre, while taking nothing from his save nothing itself: "What I wish is that this in-itself might be a sort of emanation of myself while still remaining in itself."[9]

Sartre's reiteration of the Platonic dream of mastery is sustained by the matter upholding the metaphor of sliding. The skier who cuts a path across the field of snow becomes the for-itself which enacts determinations upon the blankness of the in-itself; and as the mass of the skier and his weight upon his skis embody the choice of the for-itself, so the snow, which melts and compresses to support this weight during the instant of its passing, undergoes changes of state allowing it to register and then to annihilate the imprint which would effect a "this".[10] Mobilized by the ideal of a temporal flight leaving behind a past whose "was" figures blank facticity, skiing upholds the inviolability of the snow. However, Sartre's concern for the surface, for the superficiality of skiing, protects the integrity not merely of the snow-template but of the skier himself. It is here that the threat of *passage* is addressed and diffused: " . . . the sliding is action at a distance; it assures my mastery over the material without my needing to plunge into that material and engulf myself in it in order to overcome it."[11] The taming of water is realized in snow; water appears only in the instant of compression, or submission, marking the skiers's choice. Indeed, although Sartre would find water-skiing the quintessential activity of sliding, its "ideal limit", he celebrates the unique duality of snow: while being that which, like water, "gives support, docile and without memory', snow also "remains impenetrable and out of reach."[12] In this impenetrability and distance snow is far safer than water; although Sartre can celebrate water as the ideal matter which never records a trace, the depth of water threatens to become an upsurge that would annihilate the distance upon which mastery depends. Compared to the activity of sliding, swimming is not only obscene but terrifying.

While snow functions like Plato's "universal nature which receives," Sartre's investigation of the modes of doing and having expressed by the for-itself requires metaphors illustrating more explicit forms of possession. The universality of the field of snow becomes individualized in the attempt at appropriation which is, for Sartre, equally an attempt at identification. Sartre valorizes possession as the "*symbol* of the ideal of the for-itself or value", that which would nihilate passage in the convergence of in-itself and for-itself realized by the possessed object: " . . . the dominant term is the object possessed. . . . In possession, I am my own foundation in so far as I exist in an in-itself."[13] Sartre must emphasize the purely *symbolic* function of possession, the inconceivable ideality of the "mine", because its impossibility engenders the notion of value that it represents. It is through this symbol that value is conclusively materialized as the flank of a woman, as snow, or as tobacco, and it is here that the asymptotic surpassing towards value which propels the for-itself may assume the specificity of an *act* localized in the human body. Although Sartre states that his consumption of tobacco embodies the general desire to "possess the world", this desire is mediated not only by "a particular object" but by a particular body.[14] Indeed, smoking figures not only the physiognomy of possession but, inevitably, the physiognomy of its reversal – although such physiognomy will by finally translated by Sartre into a motion indistinct from matter itself.

Sartre's metaphor of smoking expresses the literal activity of appropriation, the absorption of matter here dissolved into vapor. Sartre states: "Across the tobacco which I was smoking was the world which was burning, which was going up in smoke, which was being reabsorbed into vapor so as to re-enter into me."[15] Yet smoking as a symbol of possession critically depends upon the cigarette, which acts as a funnel of a siphon to channel the unbounded vapor of the world. This cigarette participates as an extension of Sartre's desire, the essential instrument allowing him to suck in the smoke which would otherwise envelop him. The indispensability of this cigarette shifts his metaphor into the realm of gender, for the situation that figures the reversal of smoking turns out to be Sartre's conception of sex, where he finds the penis "devoured" and where the male penetrates only to perpetrate his own dissolution: " . . . woman appeals to a strange flesh which is to transform her into a fullness of being by penetration and dissolution."[16] Sartre's portrayal of sex finds its opposite in a metaphoric activity presumed genderless: in fact the cigarette figures a magical male sex that would suck in the matter threatening to engulf it. Situated opposite sex and reflecting its inverse, Sartre's characterization of this habit as *personal* is effaced by its resemblance to a universalized anatomy which ultimately describes not the physiognomy but the activity of matter. Such activity, whether of inhalation or of engulfment, is metaphorized in a matter figuring pure quality: "the" slimy.

Sliminess confounds the abrupt polarities embodied for Sartre in "a secret liquid quality in solids," that secret which lies intact at the heart of "the polarity of the continuous and the discontinuous, the feminine and masculine poles of the world, for which we shall subsequently see the dialectical development all the way to the quantum theory and wave machanics."[17] Sartre finds this secret revealed in the melting of snow, the medium in which water and ice retain their exclusivity while each occupying the other, as well as in quantum theory, where particle and wave can be said to exist inviolate within each other precisely because there is no articulable passage between them. The untraversable passage defines the quantum leap, the ideal disjunction – or secret – underlying Sartre's polarities: thus slime as the medium of passage poses a threat to the fundamental polarity of masculine and feminine, for-itself and in-itself.

The insidious nature of slime is realized at the moment of its incorporation into the symbolic project of appropriation. In its own matter slime *anticipates* the appropriative move of the for-itself, as Sartre states: "the slimy appears as already the outline of a fusion of the world with myself." The ideal of "mine" which might be realized in this fusion is claimed by slime's "already", and the motion with which Sartre inhaled the world is reversed: "What it teaches me about the world, that it is like a *leech sucking me . . ."*[18] Because it comes first, slime's sucking represents more than a reciprocation of Sartre's attempted possession; instead, slime undermines the terms upon which possession depends by occupying the passage between them. Sartre can only describe slime as an in-between, caught between what it might be, or, indifferently, what it was: "Slime is the agony of water. It presents itself as a phenomenon in process of becoming. . . . Nothing testifies more clearly to its ambiguous character as a 'substance between two states' than the slowness with which the slimy melts into itself."[19]

Slime is not an oscillation but an eternal becoming, a "fixed instability" unable to occupy a single state because it could indifferently occupy the other. In this condition slime can only be a "phenomenon", a quality of matter which would translate itself into pure activity. Sartre situates the becoming of slime in the word "soft", at a juncture where not only matter but the possibility of its possession is jammed at a halfway point:

> In the slimy substance which dissolves into itself there is a visible resistance, like the refusal of an individual who does not want to be annihilated in the whole of being, and at the same time softness pushed to its ultimate limit. For the *soft* is only an annihilation which is stopped halfway.[20]

This stalled annihilation threatens not the slime, whose soft surface promises the passivity of the in-itself, but the for-itself, seduced by a softness turned leech-like. Softness expresses not only the terminal instability which *is* the matter of slime, but its seductive effect upon the for-itself.

Slime's "reversal" of the for-itself's clear mastery can be expressed only through its indeterminacy. It is therefore covert, characterized by Sartre with a surreptitiousness which abruptly becomes gendered: "It is a soft, yielding action, a moist and feminine sucking . . . " The "feminine" erupts at the site of yielding-sucking to characterize not slime as a type of matter but the potential activity of *all* matter as slime: "Slime is the revenge of the In-itself. A sickly-sweet, feminine revenge . . . "[21] Once sliminess becomes a possibility inextricable from the material plenum subtending all "thises", and once its revenge becomes explicitly feminine, Sartre has not only gendered the in-itself in a submission as ideal as that of Plato's receptacle, but he has cleared a site of feminine "revenge" which can only be that of passage, or of radical ambivalence. This ambivalence represents not simply a sporadic obstacle to the for-itself's project of flight – when transposed into time, slime becomes the continuous threat of death.

Sartre projects the for-itself into time as a flight from the absolute "is" of the in-itself, the "is" which is imposed upon the for-itself as an inescapable "was": "Everything happens as if the Present were a perpetual hole in being – immediately filled up and perpetually reborn – as if the Present were a perpetual flight away from the snare of the in-itself."[22] Yet the "before" of the past would claim a more material hold upon the for-itself than that threatened by a snare, for "Like the mermaid whose body is completed in the tail of a fish, the extra-mundance For-itself is completed behind itself as a *thing in the world.*[23] This "behind" figures the space relegated to the *body* of the for-itself, that attached facticity which is a surpassed in the pursuit of value: " . . . the body, since it is surpassed, is the Part."[24] With matter situated "behind" the for-itself like the mermaid's tail and its body, the direction from which slime's sucking action will emanate is also determined. Slime can only emerge to assert the revenge of the in-itself from behind, threatening the reversal of the temporal project itself. The for-itself is haunted by the possibility that:

> . . . at the very moment when it was conscious of arriving there [it] would be slyly held back by the invisible suction of the past and would have to assist in its own slow dissolution in this past which it was fleeing. . . . The horror of the slimy is the horrible fear that time might become slimy, that facticity might . . . absorb the For-itself which *exists it.*[25]

Thus slime as the material embodiment of a sucking motion coincides with the temporal flight from the snare of matter: slime's reversal of possession is set in time as the for-itself's *re* possession by the matter of its own body. Sartre here endows slime – or the in-itself – with a vengefulness which must be fully opposed to the for-itself's project of surpassing-towards: slimy time represents the "Antivalue".[26]

With the antivalue, it becomes evident that Sartre's feminization of matter has generated an irreparable schism – between the ideal lover, the water, and the snow that nourish the choices of the for-itself, and the slime, the past, the body that reach out from themselves to engulf him. This schizoid matter seems *itself* to occupy the passage between for-itself and in-itself, casting its shadow as far back as Sartre's founding cosmogony to render the in-itself, as the mute ground of free choice, an ideal.

By inscribing all freedom upon a ground which is always already feminine and always already capable of enacting a revenge inextricably linked to the sucking motion of reversed possession, Sartre seems to have gendered the very activity of choice. Sartre's freedom is jeopardized in the face of the mastery over the feminine which subtends its very possibility. Slime, the metaphor which would close *Being and Nothingness* with a horrifying (or humorous?) anxiety, instead recoils upon Sartre's text to compromise even the founding moment of his cosmogony. By dislodging matter from the perfect self-identity which grounds the choice of the for-itself, slime trespasses beyond the space of Sartre's metaphor to enact its own slimy revenge.

In a cosmogony predicated on the integrity of determining consciousness and receptive matter, Sartre's greatest fear can be expressed as follows: "If I sink in the slimy, I feel that I am going to be lost in it; that is, that I may dissolve in the slime . . ."[27] Once the incipient revenge of the in-itself emerges as a constant threat to Sartre's vision, distance between for-itself and in-itself can be interpreted as a space of mastery. The distance sustaining the relation between consciousness and gendered matter may be metaphorized in the activity of *sight*, particularly the sight of a painting whose matter yields to its viewer's gaze. Perhaps the most obvious way of exemplifying the terms of Sartrian vision in painting is by looking at nudes, which re-enact this relation quite literally.

Alexandre Cabanel's *La Naissance de Vénus* was shown in the Paris Salon of 1863 at the same time as Edouard Manet's *Déjeuner sur l'herbe.* By representing a nude, each painting articulates a particular relation not only between viewer and canvas, but between the painter and his model. While Manet's painting will be aligned later with Maurice Merleau-Ponty's re-vision of Sartrian sight, Cabanel's Venus can be seen to embody the terms of *Being and Nothingness*. True to his painting's title, Cabanel portrays a *birth*: the putti which hover over this Venus gloat in the painter's mastery of not simply the representation, but the re-creation of her body. Sprawled before the gaze of the viewer, this Venus is unmarked by any experience which she might claim as her own. Prior lineage is effaced in the body that renounces the constraints of both particularity and anatomy, leaving Cabanel's studio model present only vanishingly. Perfectly submissive to the dissolution of her own genealogy in painterly re-creation, Cabanel's model becomes pure *quality*, her docile promise only embodied to the extent that it is given a gender. The ideal quality of the model, who bends to the will of the painter without leaving a trace of herself in his re-formation, can be characterized in his term "smooth," as evoked to describe the in-itself above:

> Smooth – it is what re-forms itself under the caress, as water reforms itself in its passage over the stone which has pierced it.

Cabanel's Venus not only elicits a gaze which would mimic the activity of Sartre's determining consciousness, but, in return, illustrates the smoothness which is its ideal object. Indeed, not only the painter's model is forgotten in the act of re-formation: prior to *any* nude body is the smoothness of pure mute matter. Yet it is *Venus* who rises from the water surrounding her to figure its quintessence; Sartre's evocation of the erotic "smooth whiteness of a woman's body" situates the ultimate submissive appeal of water in female anatomy.

Manet's *Déjeuner sur l'herbe* can be read as an allegory of the fate of the painter's model counter to that seen in Cabanel's *Naissance de Vénus*. This painting may be seen to represent the scaffolding suppressed by Cabanel's Venus: here a naked model poses as nothing other than *herself* taking a rest from posing. Although the gazes of the men flanking her trace a trajectory which excludes her completely, she stares out of Manet's canvas at the viewer to affirm her particular presence within *this* painting, resisting the right of the shadowy other painting contained in a narrative that would actually claim her as its nude. Manet's model effaces that idealized nude as surely as Cabanel's Venus reciprocally effaces its real model.

Déjeuner sur l'herbe figures the stubborn materiality of the modern model, who no longer melts into watery quality between poses, but who claims an ineluctable presence become part of a painting's address to its viewer. Her particularity defies idealization, and what would be an invisible moment between painted scenes is here staged, its own spectacle. Paradoxically, the nude is both a quintessential model, herself, and not a model at all, herself. The painting's second woman is part of the scenery framing this spectacle: merging into the matter of its flatly painted backdrop, she is an element of past convention, the sightless ghost of Cabanel's Venus.

Manet's *Déjeuner sur l'herbe* figures the revision of the Sartrian cosmogony found in Maurice Merleau-Ponty's *Phenomenology of Perception* (1945). Merleau-Ponty's book represents an attempt to bypass the distinction of for-itself and in-itself, consciousness and matter. In his effort to bridge the distance which defines the mastery of consciousness over matter, Merleau-Ponty integrates Sartre's determining gaze into a body which will not relinquish it. To situate that body at a juncture which would nihilate the space between consciousness and world, Merleau-Ponty compares it to a work of art:

> The body is to be compared, not to a physical object, but rather to a world of art. . . .
>
> A novel, poem, picture or musical work are individuals, that is, beings in which the expression is indistinguishable from the thing expressed, their meaning, accessible only through direct contact, being radiated with no change of their temporal and spatial situation. It is in this sense that our body is comparable to a work of art. It is a nexus of living meanings. . . .[28]

Merleau-Ponty's evocation of the work of art as a being "in which the expression is indistinguishable from the thing expressed" coincides with Samuel Beckett's defense of the work of James Joyce, a defense that sets the terms for an explicitly modernist perception:

> Here is direct expression – pages and pages of it. . . . Here form *is* content, content *is* form. You complain that this stuff is not written in English. It is not written at all. It is not to be read – or rather it is not only to be read. It is to be looked at and listened to. He is not writing about something: he is writing something.[29]

The "direct expression" which constitutes Joyce's modernism is equally that of Merleau-Ponty's work of art, "accessible only through direct contact" which dissolves the distinctions between reading, looking, and listening. Merleau-Ponty's implicit rejection of the classical ideal of art in favor of the materiality of modernism anticipates a body which is, as he says above, "a nexus of living meanings". Merleau-Ponty says of his experience of Cézanne, "it is the sight of the pictures which provides me with the only existing Cézanne":[30] opposed to Sartre, Merleau-Ponty's sight serves not as the metaphor for a process of abstraction or distancing but rather as an affirmation of the paintings' lived presence.

The meanings of Merleau-Ponty's body itself must be lived, not congealed like the dead weight of the Sartrian body, the extra-mundane facticity which Sartre attaches to consciousness "like the mermaid whose body is completed in the tail of a fish".[31] For Merleau-Ponty, bodily meanings are derived not simply from the body's own matter, but from the entire world: the body becomes nothing more – or nothing *less* – than a "hollow, [or] a fold"[32] through which the world lives.

Because it cannot claim the world at a distance, Merleau-Ponty's modernist body undermines the terms of Sartrian possession:

> The thing is inseparable from a person perceiving it, and can never be actually *in itself* because its articulations are those of our very existence . . . To this extent, every perception is a communication or a communion . . . a coition, so to speak, of our body with thing.[33]

Merleau-Ponty's metaphor of coition displaces the Sartrian metaphor of sight, a revision whose implications for the viewer of painting have been anticipated in the work of Manet. Instead of a painting which has been invested with transcendent meaning by the artist, a painting whose ideal meaning is transmitted to the viewer against the grain of its own matter, Merleau-Ponty sees a painting whose entire significance resides in its living "communion" with the body facing it, that material "coition" which represents Sartre's greatest antipathy. For Sartre, it is precisely such coition that is infused with fear of dissolution into feminine slime.

Manet's *Déjeuner sur l'herbe*, where the model who was submissive to Cabanel is decisively personalized to confront the viewer with her non-idealized paintedness, can be said to emblematize the qualities of modern painting. Georges Bataille, for example, celebrates Manet as the first modern painter, in whose work "the picture obliterates the text [and] the meaning of the picture is not in the text behind it but in the obliteration of that text". Bataille's emphasis on the obliteration of any external text transfers all meaning to the painted surface alone: thus for the first time in the history of art, Bataille finds in Manet painting as "a pure state of being".[34] Despite the ruthlessness of Bataille's appropriation of Manet, he does evoke the depth of Manet's rupture with preceding painting. It is these qualities, or the qualities of modern art in general, which Merleau-Ponty chooses to evoke as a metaphor for the perceiving body: the communion between an art object and the body mimics the communion between the body metaphorized as an art object and the world.

Can *modernist* art, itself resisting the duality of form and content which buttresses Sartrian vision, be seen as Merleau-Ponty's only and inevitable metaphor for the body? In *The Prose of the World*, where he discusses the relation of modern to classical painting, Merleau-Ponty elaborates a sense of modern art which further refines what are both the uprecedented possibilities and the radical aporias of his perceiving body.

While I have illustrated the distinction between classical and modern painting through the situation of the painter's model, as seen in the paintings of Cabanel and Manet, Merleau-Ponty uses the notion of perspective to trace this evolution. At the outset, he situates the representation of depth within a cultural, and not a perceptual, field: " . . . classical perspective is not a law of perceptual behavior. It derives from the cultural order, as one of the ways man has invented for projecting before himself the perceived world, and is not a copy of this world."[35] Classical perspective is a method of inclusion that would spatialize time to incorporate the successive points marked by the trajectory of the painter's gaze into a single focused image. Yet as a system of order that is always already imposed upon the painter's representation, perspective can only pronounce the death of perception:

> In spontaneous vision, I had the experience of a world of teeming, exclusive things which could be embraced only by means of a temporal cycle in which each gain is simultaneously a loss. In the other case, this world crystallizes into an ordered perspective. . . . The whole scene is in the past, in the mode of completion and eternity. Everything adopts an air of propriety and discretion. Things no longer call upon me and I am not compromised by them.[36]

Classical perspective is a means of possessing the world, a geometry which bypasses time to insist upon the eternal docility of space. Merleau-Ponty's bodily communion is denied by an order than muffles perception: "the experience of a world of teeming, exclusive things" can only be antithetical to the designs of transcendent geometry. Proclus, the neoplatonic philosopher who commented upon Euclid's *Elements*, articulates the classical antagonism between Idea and perception:

> . . . the Pythagoreans recognized that everything we call learning is remembering, not something placed in the mind from without. . . . [The soul] possesses them all [geometrical ideas] in an essential, though latent, fashion and brings each of them to light when it is set free of the hindrances that arise from sensation.[37]

Proclus reiterates the Platonic distinction between the sensible and the intelligible; and this submission of the material world to disembodied eternal form returns us to the metaphor of *sight*, as Michel Serres makes clear: " . . . in the final analysis the path [from practice the geometrical theory] consists in forsaking he sense of touch of that of sight. . . . Here, to theorize is the see."[38] The "light" of knowledge, opposed to the "hindrances" of perception, equates sight with a "remembering" or recognition that takes nothing from the material world but instead finds there what it already knows. The sight which would impose geometrical perspective upon the world must forsake its own body in a mastery over matter which Sartre elaborates in the relation between in-itself and for-itself.

In valorizing communion over Sartrian vision, Merleau-Ponty revitalizes sight itself by re-temporalizing it.[39] Renouncing the eternal fixity of classical perspective, Merleau-Ponty instead evokes the fluidity which would make not only the body but the world "a nexus of living meanings". Perception congeals into order only at the body's end; thus in terms of painting, Merleau-Ponty would find a living work to be one which is not *finished*:

> One of the grandeurs of modern thought and modern art is to have loosened the false lines which tied a valuable work to the finished work. Since perception itself is never *finished*, since it gives us a world to express and think only through those partial perspectives which it envelops and whose ineffable evidence is not the kind we possess . . . [modern art] many also be the recognition of a mode of communication which does not pass through objective evidence, a signification which does not refer to an already given object but constitutes and inaugurates its object . . .[40]

Here the body, already metaphorized as an (implicitly modern) work of art, is placed before a painting which is likewise itself not a "given object" but a lived signification. Perception is opposed to the viewer's recognition, or platonic remembrance; instead of attempting mimesis, the modern painter eludes the constraints of verisimilitude altogether, as Merleau-Ponty says: "Modern painting, like modern thought generally, absolutely obliges as to admit a truth which does not resemble things . . .[41] In a painting which is the active constitution of its object, as the body is the active constitution of the world's meanings, "truth" resides not in the painting's citation of objective evidence but in the contingent resolution loaned it by its viewer: the only painting which *can* participate in the viewer's perception must revoke its claim to exact representation. Finished painting, like a finished body, is for Merleau-Ponty no more than an element of taxonomy.

Yet the very resonance of Merleau-Ponty's comparison of art and body throws the featurelessness of both into relief. The inevitability of his metaphor is enforced by what it excludes: in the case of modern painting, the circumstances of its constitution as art; and in the case of the body, its gender, race and class. Merleau-Ponty's discussion of the experience of modern painting implicitly relies upon a standard of *quality* which he never elaborates except when he must concede that it is only "the better or more worthwhile [moderns]"[42] who evoke a full response. Indeed, that the designation of "worth" emerges from a cultural context is demonstrated in the case of Manet's *Déjeuner sur l'herbe*, a work by the painter who inaugurated the modernism that Merleau-Ponty celebrates. Yet representative criticism of the Salon des Refusés in 1863, the exhibition of officially rejected works where the painting was shown, states: "This is a young man's joke, a shameful open sore not worth exhibiting this way."[43] While Merleau-Ponty insists upon classical perspective as a cultural artifact and not a perceptual intuition, he exempts modernism from a parallel investigation despite the fact that the first viewers of Manet's work were unable to "communicate" with it.

Just as conventions of value are obscured by the universal effect of modern painting, the gender of Merleau-Ponty's body is lost in the fluidity of the experience which precludes its claiming a static essence. Yet while the universality of modernism bypasses the historical possibility that certain subjects were excluded from the very constitution of the term, so the universality of Merleau-Ponty's perceiving body does not recognize the historical priority of the gender which would threaten to appropriate it as always already male. Merleau-Ponty's genderless body is both a figure of radical liberation, freed from the constraints of essential sexuality and fully incorporated into the living matter of the world, and a figure of dubious antecedents which Merleau-Ponty makes no attempt to acknowledge or explicitly refute.

Judith Butler, in her analysis of Merleau-Ponty's brief discussion of the sexual body in the *Phenomenology of Perception*, demonstrates that the desiring body described in this chapter can only be male. Schneider, Merleau-Ponty's defective case study, is deficient precisely because he cannot respond to woman's bodies *in general*, as Merleau-Ponty observes: "for Schneider a woman's body has no particular essence".[44] Butler finds here that women, reduced to fixed objects of desire, are denied the existential status of Merleau-Ponty's "universal" body. In this universality Butler finds no acknowledgement of the heterosexual order which would define both Schneider's ailment and the "particular essence" of woman:

> Strangely enough, Merleau-Ponty's effort to describe lived experience appeals to an abstract metaphysical structure devoid of explicit cultural reference, and yet once this metaphysical structure is properly contextualized as the cultural construction of heterosexuality, we do, in fact, seem to be in the presence of a widely experienced phenomenon. . . . We are left with a metaphysical obfuscation of sexual experience, while the relations of domination and submission that we do live remain unacknowledged.[45]

Yet perhaps Monique Wittig's evocation of a similarly unmarked corporeal state indicates a strategy by which this phenomenological body may be appropriated to envision a radical subversion of the very order that remains unacknowledged by Merleau-Ponty.[46]

Despite its abstraction from historical circumstance, Merleau-Ponty's modernist body is a figure of enormous potential for appropriation and revision. While he does not acknowledge the omissions underpinning the body which he posits as universal, that body remains radically incompatible with the Sartrian division of the world into being and nothingness, into the determining for-itself and the feminine in-itself which can only submit to its designs from a distance. Like the modernist art which can be characterized by its aliveness, so Merleau-Ponty's staunchly material body is a figure of endless potential because it – and its world – resists the Platonic recognition, the Sartrian nothingness, which would already know it and shape it from a distance.

Duke University

NOTES

1 Jean-Paul Sartre, *Being and Nothingness* (New York: Washington Square Press, 1966), trans. Hazel E. Barnes, p. 48.
2 *Ibid*., p. 49.
3 *Ibid*., p. 121.
4 *Ibid*., p. 296.
5 *Timaeus* (Macmillan Publishing Co., 1985), trans. Benjamin Jowett, 50a.
6 Sartre, *Being and Nothingness*, p. 792.
7 *Ibid*., p. 515.
8 *Ibid*., p. 740.
9 *Ibid*., p. 743.
10 In fact, Sartre does concede that "Sliding on snow is already less than perfect; there is a trace behind me by which I am compromised, however light it may be." (p. 746.) However, he minimizes the effect of this trace: "Besides when we let ourselves slide down

the slope, we are accustomed to the illusion of not making any mark; we ask the snow to behave like the water which secretly it is." (p. 746.) Although the metaphor of skiing is troubled by the faint presence of tracks, snow seems ultimately to be a matter more easily appropriated than water, which threatens to engulf the water-skier who loses his balance.

[11] Sartre, *Being and Nothingness*, p. 745.
[12] *Ibid.*, p. 746.
[13] *Ibid.*, p. 755.
[14] *Ibid.*, p. 762.
[15] *Ibid.*, p. 761.
[16] *Ibid.*, p. 782.
[17] *Ibid.*, p. 766.
[18] *Ibid.*, p. 773.
[19] *Ibid.*, p. 774.
[20] *Ibid.*, p. 775.
[21] *Ibid.*, p. 777.
[22] *Ibid.*, p. 208.
[23] *Ibid.*, p. 208.
[24] *Ibid.*, p. 429.
[25] *Ibid.*, p. 778.
[26] *Ibid.*, p. 778.
[27] *Ibid.*, p. 777.
[28] Maurice Merleau-Ponty, *Phenomenology of Perception* (London: Routledge, 1962), trans. Colin Smith, pp. 150–151.
[29] Samuel Beckett, in "Dante . . . Bruno. Vico . . . Joyce," *Transition*, no. 16–17, June 1929, p. 248.
[30] Merleau-Ponty, *Phenomenology of Perception*, p. 150.
[31] *Sartre, Being and Nothingness*, p. 208.
[32] Merleau-Ponty, *Phenomenology of Perception*, p. 215.
[33] *Ibid.*, p. 320.
[34] Georges Bataille, *Manet* (Geneva: Editions d'Art Skira S.A., 1983), trans. A. Wainhouse and J. Emmons, pp. 62–3.
[35] Maurice Merleau-Ponty, *The Prose of the World* (Evanston, IL: Northwestern Univ. Press, 1973), trans. John O'Neill, p. 51.
[36] *Ibid.*, p. 53.
[37] Proclus, *A Commentary on the First Book of Euclid's Elements* (Princeton: Princeton Univ. Press, 1970), trans. Glenn R. Morrow, p. 37.
[38] Michel Serres, "Mathematics and Philosophy: What Thales Saw . . ." in *Hermes: Literature, Science, Philosophy* (Baltimore: Johns Hopkins Univ. Press, 1982), trans. Josué V. Harari and David F. Bell, p. 86.
[39] In *A l'ombre des jeunes filles en fleurs*, Marcel's first vision of "la petite bande" of girls at the beach of Balbec illustrates the "fluid" perception that Merleau-Ponty (and Bergson) celebrates: ". . . je voyais émerger un ovale blanc, des yeux noirs, des yeux verts, je ne savais pas si c'était les mêmes qui m'avaient déjà apporté du charme tout à l'heure, je ne pourvais pas les rapporter à telle jeune fille que j'eusse séparée des autres et reconnue. Et cette absence, dans ma vision, des démarcations que j'établirais bientôt entre elles,

propageait à travers leur groupe un flottement harmonieux, la translation continue d'une beauté fluide, collective et mobile." (Paris: Gallimard, 1988, p. 356.) The brief absence of spatial demarcations lends "la petite bande" a beauty which moves in *time*, "la translation continue" of Marcel's own evolving vision.

40 Merleau-Ponty, *The Prose of the World*, p. 56.

41 *Ibid.*, p. 66.

42 *Ibid.*, p. 55.

43 Louis Etienne, quoted in George Hamilton Heard, *Manet and his Critics* (New Haven: Yale Univ. Press, 1986), p. 45.

44 Merleau-Ponty, *Phenomenology of Perception*, p. 156.

45 Judith Butler, "Sexual Ideology and Phenomenological Description," in *The Thinking Muse: Feminism and Modern French Philosophy*, ed. Jeffner Allen (Bloomington: Indiana Univ. Press, 1989), p. 97.

46 Here I am thinking in particular of *The Lesbian Body*, as well as the essays, such as "On the Social Contract," in which Wittig relies upon the concept of some kind of pre- or post-social corporeal space.

GEORGE B. MOORE

PARANOIA AND THE AESTHETICS OF CHAOS IN THOMAS PYNCHON'S *GRAVITY'S RAINBOW*

What many are now calling "the new science of chaos" is actually a diverse group of scientific investigations that promote a view of the world that corresponds, not accidently, with various textual strategies that have evolved over the last few decades.[1] The similarity of these visions has lead some to speculate on the line (i.e., linearity) of influence. The argument generally runs that science has given birth to new insights in the structures of the world that are reflected in textual production and interpretation, or, the inverse, that literature has anticipated certain discoveries in science. Neither position, however, seems to take into account the complex dynamics of the cultural exchange of ideas between the two fields. But more importantly, neither argument takes into account the fact that both science and literature, to some extent, construct the contexts for their own interpretations. An awareness of this constructive factor might take us in a new direction: toward the metafictional nature of the "new science", making us realize a scientific reflexivity based in the awareness of its own power of reorientation. In the end, it appears that science too promotes the paradoxical nature of language (even as literature does) by suggesting a chaos that is anything but chaotic.

It is the extensive field of similarities and differences in textual interpretation and text-production that suggests both the benefits and dangers of this new approach to literary interpretation. Kenneth J. Knoespel, in a recent essay, "The Emplotment of Chaos: Instability and Narrative Order", addresses the problem of the analogy between science and literature by arguing that both are processes of mapping.[2] He warns that when we compare critical ideas such as deconstruction and information theory we may be too eager to produce a theoretical commonality, and therefore tend to overlook the conceptual nature of the language in both areas. Knoespel reminds us that, "Whether the scientist maps the earth's surface, microorganisms within its geological structure, or the gaseous clouds of a supernova, the process involves the ongoing inscription of information."[3] As an alternative to reading chaos in critical theory, Knoespel would have us think in terms of how a critic, such as Jacques Derrida, would read the new order suggested

M. Kronegger and A-T. Tymieniecka (eds.), Analecta Husserliana XLII, pp. 203–218.

by non-linear dynamics, that is, with an eye for the reinscription of scientific authority. "Deconstruction challenges the impulse to build and overarching system found in the very inclination to view chaos theory as the entrance to a new scientific orientation."[4] Although both systems have their own nomenclatures, critical theory would have us recognize the inscription of authority as the condition of our inestigations in both areas. Knoespel does not, however, abandon the notion of analogy. Instead, he suggests that we view the connection between chaos and narrative theory in terms of local sites of interpretation: "The emplotment of chaos arises less from universal expectations than from the demonstration of its applicability to a range of local problems."[5]

Knoespel's cautious reaction to the popular call for a new revolution in scientific thinking seems to be shared by Ilya Prigogine and Isabelle Stengers in their key work on non-linear dynamics and complex systems, *Order Out of Chaos: Man's New Dialogue with Nature*.[6] These authors stress the point that science has not so much diverged from its historical course as incorporated some old and new perspectives. Rather than disproving the science of the past with startling new discovers, the authors suggest that order out of chaos is a natural extension of the thinking that produced quantum physics. In the conclusion to their work, Prigogine and Stengers remind us:

> It is important to point out that the new scientific developments we have described, the incorporation of irreversibility into physics, is not to be seen as some kind of "revelation," the possession of which would set its possessor apart from the cultural world he lives in. On the contrary, this development clearly reflects both the internal logic of science and the cultural and social context of our time.[7]

Literature, of course, is a reflection of "the cultural and social context of our time". And the fact that time has returned to haunt the processes of science cannot help but establish a kinship with the time-bound structures of narrative fiction and narrative theory.

It is this cultural exchange and its accompanying problems of analogy that bring us to the work of Thomas Pynchon. Pynchon raises many of the same questions of scientific orientation as chaos theorists, yet he continues to displace the authority of these theories by exposing elements of contrariety and paradox. In a critical sense, Pynchon deconstructs the notion of science as absolute authority by showing how it reduces human experience to unresolvable contradiction. The collision of the human and the scientific in *Gravity's Rainbow* is the site of twentieth-century physics's recognition that processes are indeed time-bound, and that order cannot be found in the world by imposition, but rather discovered only in the flux of processes always involved in the dynamics of chance and change.

The key to this reading of Pynchon's work is his use of paranoia as a narrative strategy. *Gravity's Rainbow* is a textual labyrinth of subplots and minor characters, all interwoven into a narrative structure that's only coherence comes from the notion that each action and event is dependent upon a network of other events, all of which are either secretly planned or strikingly coincidental, and the question of which is never resolved. The novel is held together by the single shared perspective that the world is made up of connections: scientific, social, political, economic, and that to recognize them at the level of the network itself is to find oneself, as reader or as character, involved in one way or another in the aesthetics of chaos.

Pynchon makes an important connection when he links the concept of entropy to the human experience of paranoia. Both are ways of relating to the world and universe as closed systems: the paranoid imagines connections between all events, or recognizes actual links, while entropy is the condition of the movement of universal events toward a single state of equilibrium. In classical physics, entropy is the measure of the amount of energy unavailable in a system that is undergoing change; that is, changes within the system cause a decrease in energy and an increase of entropy. Because there is always a certain amount of energy lost in the process of transformation – but specifically, within a closed system – entropy has come to be known as "time's arrow", the factor of irreversibility in universal processes. Pynchon and others relate this to the devolution of the universe itself through slow decay. In pre-chaos theories of universal energy exchange, entropy is generally associated with the movement of the universe toward equilibrium or heat death. Pynchon uses this notion of entropy to suggest the relationship between human endeavors and scientific order: as science struggles to understand and thereby control both natural forces and human behavior, it reduces the exchange of ideas and human interactions to known factors, to predictions and foreseeable procedures, to controlled behavior. This regime of human hyper-order, for Pynchon, is a form of equilibrium, a final stage where all things are controlled and nothing is left to chance. The paranoid not only sees this connection, but recognizes the omnipresence of the system that produces it, giving rise to the idea of worldly order that represents scientific authority.

The key example from the book is Tyrone Slothrop, a lieutenant in war-torn London whose "rocket-dowsing" erections connect him to the V2 blitz and to the world of ballistics and probability factors. Slothrop's importance as the central narrative figure is reinforced by his connection to the missile subculture that interlocks all the various subplots and minor characters. Slothrop's erections map the incoming V2s by corresponding to his sexual exploits around London, at least this is what we are initially led to believe. Later evidence suggests that at least some of these exploits, mapped on the wall of his cubical, are imaginary, or uncertain, throwing open the question of cause and effect and suggesting that probability patterns can form out of chance encounters. We are given enough information to assume that Slothrop's erections somehow sense the pattern of V2 hits before they arrive, but we are also given information that leads us to doubt this assumption, information that throws the question of Slothrop's involvement open to speculation. Brian McHale uses this condition in the text, what he calls "the ontological instability", to warn us against the "temptation to apply . . . naturalization on a large scale".[8] According to McHale, a number of critics have attempted to reduce the novel to the product of its own narrative suggestions, only to find that the patterns in *Gravity's Rainbow* are never more than probability patterns and are in no way limited by the sanctions of fictional reality or verisimilitude. In relation to chaos theory, Prigogine and Stengers suggest that science now must deal with such conditions of order rising out of chaos, and that the same dependency on cause and effect has limited the conventional scientific view of the universe: "Our universe has a pluralistic, complex character. Structures may disappear, but also they may appear. Some processes are, as far as we know, well described by deterministic equations, but others involve probabilistic processes."[9] *Gravity's Rainbow* plays out the drama of scientific determinism as it reaches the limits of its credibility, and, at those limits, we find new orders rising out of the disarray of narrative suggestions.

In our efforts to achieve an overview of relationships like Slothrop's in the novel – if indeed an overview in the traditional sense is possible – we find ourselves steeped in an unusual sense of textual paranoia, an almost unconscious insight that somehow all these things are indeed connected if we can but decipher the Pynchonian semiotics. At the same time, we find that paranoia is not the singular domain of the reader grappling with Pynchon's labyrinth; most of the major characters suffer from this double-edge desire/fear to make patterns out of the disorder that surrounds them. In Slothrop's case, paranoia takes the form of his own search for connections on two primary levels: the first is constituted by his efforts to discover the truth about his erections and their relationship to the Rocket; the second is the Weltanschauung that evolves out of the myriad clues that lead him toward that truth. But the idea of truth hidden within this scheme is still one of cause and effect, a product of conventional scientific thinking. As Slothrop's nemesis, Pointsman, suggests, "He's out there, and he can *feel them coming* days in advance. But its a reflex. A reflex to something that's in the air *right now*. Something we're too coarsely put together to sense – but *Slothrop can*.[10] Trapped in causal relationships established by others, Slothrop's general tendency is to follow the patterns that he finds himself apart of, even when he has no hint as to their eventual outcomes.

Along with the pervasiveness of self-similar events and chance encounters comes the eventual break down of strict causal relationships. We are finally forced, as readers, to consider paranoia in its ontological associations, as a perception of the world that we cannot delineated so much as sense among the various levels of textual involvement. Peter Stoicheff, in his recent work, "The Chaos of Metafiction", captures this sense of the production of order in a metafictional text like Pynchon's:

> Metafiction . . . sustains a complexity which is initially experienced by the reader as disorder or noise, incapable of reduction to authorial intention or ordered message. To "understand" a metafictional text, one must reject seeing it as a vertical organization of a text's components into a closed order that is interpreted as meaning. Rather, one replaces this view with the recognition of lateral patterns in which disorder becomes order, mystery becomes illuminations and then fragments into a new order.[11]

This sense of the movement of chaos into order is what Slothrop experiences as the complexities of his quest coalesce into new structures of pervasive interconnectedness. For instance, at one point when he is aided in his search for the truth, he thinks to asks "Sa-a-a-ay . . . Why are all you folks helping me like this? For free and all?," we're told: "Who knows? We have to play the patterns. There must be a pattern you're in right now."[12] The critic Antonio Marquez reinforces the multilevel phenomenon of paranoid awareness here when he suggests that "underlying Pynchon's expansive treatment of paranoia, which is both serious and comic, lie epistemological and ontological questions: what sense can be made of a shifting, disoriented, and multidimensional world?"[13] Here, the world and the text become symptoms of a single paranoid interrelatedness, while this shift to the metaphysical is shared by the reader and characters alike. Many of the main characters experience connections on a level that suggests their involvement in the creation of patterns. This occurs as a continual process toward epistemological uncertainty: "Those like Slothrop, with the greatest interest in discovering the truth, were thrown back on dreams, psychic flashes, omens, cryptographies, drug-epistemologies, all dancing on a ground of terror, contradiction, absurdity."[14] If paranoia is an awareness of interconnectedness, it is not restricted to verifiable phenomena, but manifests itself in conscious and unconscious associations as well as on the level of empirical phenomena.

In *Gravity's Rainbow*, the paranoid's view of order is more than a metaphor for the spontaneous self-organization of chaotic systems. Phynchon confronts the limits of classical science found in Newtonian physics, and by constructing and then synthesizing self-similar patterns in the novel, he shows how a new sense of order can evolve out of the disruptions to simple cause and effect systems. In 1865, Friedrich August Kekule von Stradonitz dreams of a serpent holdings its own tail and thereby discovers the molecular structure of benzene. This dream begins a cycle of events that eventual lead to the establishment of the plastics industry, one of the symbols in the novel of artificial control. It is not Kekule's dream in itself that represents scientific control, but rather its application; Pynchon would have us believe that it is Kekule's interpretation of his dream that subverts the natural order suggested in the dream itself. The narrator complains of "the meanness, the cynicism with which this dream is to be used. The Serpent that announces, 'The World is a closed thing, cyclical, resonant, eternally-returning,' is to be delivered into a system whose only aim is to *violate* the Cycle." But by subverting the dream of the serpent, the System also invites its own destruction: "The System may or may not understand that it's only buying time. And that time is an artificial resource to begin with, of no value to anyone or anything but the System, which sooner or later must crash to its death . . . ".[15] A return to natural order, through the processes of decay and death, suggests a pattern that goes against the scientific appropriations of Kekule's dream for the creation of artificial substances like plastic. The new patterns are linked with the old, promoting a view of the world that focuses on the complexities and dynamics of renewable processes:

> Who sent this new serpent to our ruinous garden, already too fouled, too crowded to qualify as any locus of innocence – unless innocence be our age's neutral, our silent passing into the machineries of indifference – something that Kekule's Serpent had come to – not to destroy, but to define to us the loss of . . .[16]

In the awarenesses of its characters, *Gravity's Rainbow* does not often go beyond the suggestion that alterative patterns exist; it focuses instead on the self-destruction generated by the systems that support scientific reduction. New orders are often tied to natural processes, to human potentials, and in the novel they rise up as a disorganized but well-meaning "counterforce" to the machinations of scientific control.

Returning to Slothrop, we see that his position might be read as the unifying thread of the paranoids' conditioning within the text, but his paranoia is only one dimension of the collective pattern of self-organizing elements. Other patterns arise among characters involved in the reconstitution of order out of the fragmentation of old systems. In addition, we find a host of individual elements in the novel that represent reversals of expectations, of normalized linear progressions and of causal relationships. All of these various elements test the limits established by classical scientific thinking. Consequently, the reader discovers that the act of reading is itself a pattern-forming process, and that our involvement in the text demands that we lend ourselves to a new sense of narrative structure and cognitive process. As Prigogine and Stengers suggest:

> As randomness, complexity, and irreversibility enter into physics as objects of positive knowledge, we are moving away from this rather naive assumption of a direct connection between our description of the world and the world itself. . . . physicists had to introduce new mathematical tools that make the relation between perception and interpretation more complex. Whatever reality may mean, it always corresponds to a an active intellectual construction.[17]

The theoretical limits established by the ideas of entropy are not, after all, modeled on the universe, but on closed systems. The universe retains its elements of the unknown, the timeless, and in its entirety, whatever that may encompass, rejects, in the complexity of its interrelations, the simplicity of isolated and controlled environments.

Pynchon seems to recognize this difference between macrocosmic processes and the microcosmic models, models that over the centuries we have mistaken for the universe itself. In *Gravity's Rainbow* we find the break down of classical scientific notions coinciding with the birth of new orders that run counter to the persistent deterioration of the universe through entropy. One of the most important examples of this is the Zone-Hereros, a groups of black Africans who make up one of the German V2 batteries, and who were originally being groomed by the Third Reich for the conquest of African. Again, it is possible to see how paranoia plays a part in the new formulations of order. The Hereros are torn between two possibilities as the war ends: drawn to the idea of racial suicide, to a return to the zero: "They calculate no cycles, no returns, they are in love with the glamour of a whole people's suicide – the pose, the stoicism, and the bravery." At the same time, within the group are those who see the possibility of a return to tribal unity, where they " . . . will not be bound, like the Rocket, to time. The people will find the Center again, the Center without time, the journey without hysteresis, where every departure is a return to the same place, the only place. . . . "[18] In the character of Enzian, the Zone Hereros' leader, Pynchon has created another example of paranoid counterforce. Enzian is caught in the pull of these two forces of self-destruction and self-preservation. Drawn into the technological subculture of the Rocket, he sees clearly the entropic end of the order created by the German High Command:

It began when Weissman brought him to Europe; a discovery that love, among these men, once past the simple feel and orgazming of it, had to do with masculine technologies, with contracts, with winning and losing. . . . Beyond simple steel erection, the Rocket was an entire system *won*, away from the feminine darkness, held against the entropies of loveable but scatterbrained Mother Nature.[19]

Enzian's paranoia grows as he recognizes the spread of controlling factors begun with the German V2 campaign and now somehow infiltrating the chaos of the post-war German Zone. Like Slothrop's metaphysical musings, Enzian's move from the known universe of cause and effect relationships to the possibilities of endless chains of interacting systems creates a process that at first reduces the probabilities of chance only to force a sudden awareness of new patterns. Isolated even among the other remnants of V2 batteries, Enzian fears: "Perhaps it's theater, but they *seem* no longer to be Allies . . . though the history they have invented for themselves conditions us to *expect* 'postwar rivalries', when in fact it may all be a giant cartel including winners and losers both . . . " And as his sense of these connections grows, he begins to see how the whole system depends on minute changes that occur each second: "It comes down to this day-to-day knitting and unraveling, minor successes, minor defeats. Thousands of details, any one of which carries the chance of a fatal mistake . . . " Enzian's patterns are constructed to answer his own sense of loss of control, of increase in the entropy of the system he finds himself apart of: " . . . it is *their* time, *their* space, and he still expects, naively, outcomes the white continuum grew past hoping for centuries ago," and too, "He has the odd feeling, in moments of reverie or honest despair, that he is speaking lines prepared somewhere far away (not far away in space, but in levels of power), and that his decisions are not his own at all . . . " Eventually, his thoughts converge with what we might consider Pynchon's own position on the machineries behind war: "It means this War was never political of all, the politics was all theatre, all just to keep the people distracted . . . secretly, it was being dictated instead by the needs of technology . . . by a conspiracy between human beings and techniques, by something that needs the energy-burst of war. . . ."[20]

Out of this closed system, Enzian creates a new order. He sees the Rocket as Text, "Its symmetries, its latencies, the *cuteness* of it enchanted and seduced us while the real Text persisted, somewhere else, in its darkness, our darkness . . . " This is another order hidden as text within the order established by science, and it displaces the text of the classical configurations of cause and effect, of closed systems. His insight leads him to believe that other patterns must evolve out of the destruction of Their control, new images of natural order based on the input of chance:

> The Zone, again, has just changed, and they are already on, into the new Zone. . . . And if it should prove not to be the Rocket . . . Why then he'll have to go on won't he, on to something else . . . and if he dies before they find the True Text to study, then there'll have to be machinery for others to carry it on . . .

When we last see Enzian, he is envisioning another kind of order, a return that links science back into the time-bound processes of the earth itself: "Somewhere, among the wastes of the World, is the key that will bring us back, restore us to our Earth and to our freedom."[21]

Pynchon uses characters like Slothrop and Enzian as paranoid subjects caught in the structural paradoxes of closed systems, yet these characters eventually find ways of recognizing new relationships for themselves within the text. In the final pages, Slothrop's character dissipates, breaking up into fragments, echoes, rumors and parodies of himself so that the reader can never conclude, within the context of Their system, his eventual outcome. Enzian's character also "reconnects" in a new way, returns to a human vision that works against the equalizing control of forces in the Zone. In addition, there are complimentary situations among these characters that suggest the larger picture of negentropic designs. The primary example is the Rocket itself, which reverses the expectations of a causal relationship by traveling faster than the sound of its own approach, you can only hear the V2s if you're not at the point of impact, "if you're dead, you *don't* hear them". But Pynchon makes a point of tells us that this is not a reversal of natural law, but rather a moment of reversability or negentropy within the movement of irreversible time. The narrator says: "But the reality is not reversible. Each firebloom, followed by blast then by sound of arrival, is a mockery (how can it not be deliberate?) of the reversible process . . . "[22]

In a similar reversal of expectations, we learn that Enzian's Herero Rocket battery, known as the Schwarzkommando, are actually the creation of Myron Grunton and the propaganda wing of "The White Visitation". The narrator tells us that: "The scheme is called Operation Black Wing. My what a careful constructed five years in the making. No one could claim it all as his own, not even Grunton. It was general Eisenhower who laid down the controlling guideline, the 'strategy of truth' idea . . . " But then, within a few paragraphs, we learn that "Pirate Prentice of the S.O.E. came back with the first hard intelligence that there were indeed in Germany real Africans, Hereros, ex-collonials from South-West Africa, somehow active in the secret-weapons program."[23] The "strategy of truth" produces its own paradoxes, even going so far as to suggest that "No one [at the White Visitation] is sure who suggested the name 'Schwarzkommando' . . . ," a name, we find out later, that the Hereros have chosen for themselves. Examples like these abound in the novel, each with the singular effect of displacing the traditional narrative sense of determinism with elements of coincidence. New associations correspond with ideas of self-organization which we find at the edge of cause and effect structures represented by references to both history and science. Consequently, the novel cannot be read as a linear progression of chronological or sequential associations. As Peter L. Cooper suggests: "Seriality characterizes the cosmos and the form of *Gravity's Rainbow*: analogous events or parallel developments follow one another in loose, associational rather than logical sequence and are linked to one another not mechanically but through mutual participation . . . "[24] Effects are not linked to causes so much as to other effects, and similarities in events and ideas override traditional narrative structures based on chronology and historical sequence.

Yet this sort of shift from temporal to spatial patterning can be misleading. Although many of these examples may at first appear metaphoric, we should be careful not to limit the scope of their similarities to simple analogies. In drawing a set of guidelines for the discussion of Pynchon's work in relation to science, we must return to a common ground established by the range of theoretical and popular ideas. Many of the similarities come from Pynchon's intuitive sense of the deeper structures that govern universal occurrences, and of the connection of those occurrences to human interactions, a sense that grows perhaps out of his fascination with the paradoxes in scientific theory created in the wake of quantum physics. The key connection, then, between Pynchon and the chaos theorists is this common philosophical orientation, the history of which provides the clearest understanding of his position in terms of twentieth-century scientific thought.

Prigogine and Stengers argue that "dissipative structures", their term for self-organizing chaotic systems, must be viewed within the context of scientific developments; and in *Order Out of Chaos* they find this connection in the work of Alfred North Whitehead. They suggest that Whitehead "sought to understand human experience as a process belonging to nature, as physical existence", and to reject what he felt was the seventeenth century's inception of "scientific materialism". Whitehead's importance for Prigogine and Stengers is that he seeks to establish a connection between the macroscopic world of human existence and the microscopic, and reversible, world of classical scientific theory. According to these authors, Whitehead sees "no basic contradiction between science and philosophy. His purpose was to define the conceptual field within which the problem of human experience and physical processes could be dealt with consistently and to determine the conditions under which the problem could be solved."[25] The macrocosmic and microcosmic worlds need not be thought of differently, an idea which corresponds with Prigogine and Stengers's understanding that irreversibility, the entropic arrow of time, can incorporate the idea of spontaneous self-organization of negentropic moments:

> We now know that far from equilibrium, new types of structures may originate spontaneously. In far-from-equilibrium conditions we may have transformation from disorder, from thermal chaos, into order. New dynamics states of matter may originate, states that reflect the interaction of a given system with its surroundings.[26]

Two prominent Pynchon critics have seen the connection to Whitehead as a key factor in understanding the organization of *Gravity's Rainbow*. Antonio Marquez borrows from Joseph Slade's idea that Whitehead represents a move from cause and effect thinking to the basic principles of interactions in complex systems. Slade suggests: "Behind Pynchon's crypticism is a holistic view of the world roughly analogous to Alfred North Whitehead's concept of progress. Nowadays it is called theory of systems, and has led to a discipline which studies the energy which binds the parts of a system into a whole."[27] Moreover, Marquez makes an important connection between Whitehead's work in systems theory and "the epistemological and metaphysical basis" of Pynchon's use of paranoia. He suggests that Pynchon is using an understanding of this concept of systems interaction and complex dynamics when he discusses "creative paranoia" in the novel. "Creative or positive paranoia can counter inimical forces and form patterns or systems that are beneficent."[28] In the novel, the character of Pirate Prentice establishes the idea of "creative paranoia":

Of course a well-developed They-system is necessary – but it's only half the story. For every They there ought to be a We. In our case there is. Creative paranoia means developing at least as thorough a We-system as a They-system – . . . It's the *system* that matters. How the data arrange themselves inside it. Some are consistent, other fall apart.[29]

The emphasis on systems, on the arrangement of data, reinforces the idea of process-orientation found in Whitehead's theories and corresponds with Prigogine and Stengers's notion of the evolution of complex systems.

Prigogine and Stengers's dissipative structures and Pynchon's sense of creative paranoia share similarities on a number of levels. For all three of these authors the idea of chance and order are reinvested back into the time-bound structures of a dynamic universe, and they are linked perhaps by the history of the idea itself: " . . . for Whitehead", Prigogine and Stengers suggest, "the task of philosophy was to reconcile permanence and change, to conceive of things as processes, to demonstrate that becoming forms entities, individual identities that are born and die."[30] In Pynchon, we find Slothrop as the greatest example of this movement to reconcile permanence and change. As protagonist he succumbs to the permanence suggested by the forces of control, the scientific communities' need to define, regulate and predict the physical processes that make up the world; and yet, as human paradigm, Slothrop cannot be made an object that follows the simple patterns of universal laws. If he is part of the permanence of the text, he is also part of its existence as renewable process. Being directly connected to the larger processes that govern the novel, he cannot be reduced to a conventional product of textual representation; instead, his character is inexorably linked with the process of textualization itself. In the final pages, when Slothrop's character begins "to thin, to scatter" through the text, we learn from Kurt Mondaugen that "'Personal density . . . ' is directly proportional to temporal bandwidth . . . Temporal Bandwidth is the width of your present, your *now*. . . . The more you dwell in the past and in the future, the thicker your bandwidth, the more solid your persona. But the narrower your sense of Now, the more tenuous you are."[31] Slothrop cannot sustain that same sense of scientific authority that would establish his identity, he slips, instead, into fragments of textual representation, and into a flux of narrative elements that cannot be reduced to a single coherent meaning.

In our final analysis, we must become aware of some overriding obstacles to the analogy of the sciences with narrative fiction, and be careful not to equate the processes of interpretation and investigation in the two fields. We must at last see the nature of the similarities in the context of substantive differences. The problems generated by the types of analogies suggested here are complex and numerous, but two are perhaps most important and demonstrate the difficulties that continue to plague critics who would use scientific models for the explication of literary texts. Prigogine and Stengers caution us:

> It is not surprising that the entropy metaphor has tempted a number of writers dealing with social or economic problems. Obviously here we have to be careful; human beings are not dynamic objects, and the transition to thermodynamics cannot be formulated as a selection principle maintained by dynamics. On the human level irreversibility is a more fundamental concept, which is for us inseparable from the meaning of our existence.[32]

On the one hand, then, entropy may always remain a simple metaphor for universal decay, yet it is too early to say whether "order out of chaos" will be tied to a similar fate. The second reason suggests the complexity of this situation. Entropy here is a metaphor because the nature of the language we must use to discuss it is itself metonymic and displaces the subjects of the comparison even as it tries to bring them under the banner of a single identity. N. Katherine Hayles, in her introduction to *Chaos and Order: Complex Dynamics in Literature and Science*, certainly the most important new work in this area of comparative theory, suggests that science and literature may both be problematized by the medium through which we are force to understand them. Citing Thomas Weissert's essay in her collection, she warns us:

> . . . a phase space mapping is essentially a spatialization of a system's temporal flow. Thus complex evolutions through time are transformed into complex physical shapes that can be intuitively appreciated. I emphasize the *intuitive* aspect of this knowledge (as does Weissert) because the forms are so complex that they never resolve into completely ordered structures.[33]

Hayles seems to be suggesting that the theoretical scope of order out of chaos should not be reinscribed within the already limited dimensions of a mechanistic world view, nor can we forget that both science and literature are involved in the production and organization of signs. But literary representation is not an object of natural science; it is always already a human and artificial construct. If the inscription of information is our link between these fields, the differences in the conceptualizations of language should warn us against equating scientific investigation with text production.

We depend on language as our access to order in science as well as in literature, but literature is a world already created by signs, it forms worlds in its own likeness. It is true that science interprets the world according to its own tradition and through conventions established by the history of its own language. And when we look at that history, we see that chaos has perhaps opened doors even here; by decentralizing the Newtonian concept of physical processes, by suggesting a return to a more complex interaction of forces and systems in the universe, and by linking our ability to describe and interpret to the very structures that we investigate, chaos theory has questioned the foundations of our relationship to the world. In this, it manifests a sense of reflexivity shared by postmodern literary experimentation. In discovering new ideas of order that lie within the realms of chaos, these two disciplines share a concern with the strictures placed upon our world by our need to interpret it. This is the unspoken that haunts the words we must use to make meaning. Pynchon would appreciate this silence that invades the processes of comparison even as we try at last to bring the physical world into the sphere of the living dynamics of human understanding. We might at last turn to Enzian, caught in the chaotic Zone of post-war Europe, as a voice closest perhaps to his author's sense of this paradox: "There may be no god, but there is a pattern: names by themselves may have no magic, but the *act* of naming, the physical utterance, obeys the pattern."[34]

University of Colorado

NOTES

[1] The phrase seems to have attained it widest circulation with the publication of James Gleick's book, *Chaos: Making a New Science* (New York: Viking, 1987).
[2] Knoespel, Kenneth J., "The Emplotment of Chaos: Instability and Narrative Order," in *Chaos and Order: Complex Dynamics in Literature and Science*, ed. Hayles, N. Katherine (University of Chicago Press: Chicago, 1991).
[3] *Ibid.*, p. 103.
[4] *Ibid.*, p. 103.
[5] *Ibid.*, p. 102.
[6] Prigogine, Ilya and Stengers, Isabelle, *Order Out of Chaos: Man's New Dialogue with Nature* (Bantam: New York, 1984).
[7] *Ibid.*, p. 309.
[8] McHale, Brian, "Modernist Reading, Post-Modern Text: The Case of *Gravity's Rainbow*," *Poetics Today* (Vol. 1: 1–2, 1979), p. 97, 98.
[9] Prigogine and Stengers, 9.
[10] Pynchon, Thomas, *Gravity's Rainbow* (Viking: New York, 1973), p. 49.
[11] Stoicheff, Peter, "The Chaos of Metafiction" in *Chaos and Order: Complex Dynamics in Literature and Science*, p. 93.
[12] Pynchon, p. 257.
[13] Marquez, Antonio, "Everything is Connected: Paranoia in *Gravity's Rainbow*," in *Perspectives on Contemporary Literature* (Vol. 9, p. 92–104, 1983), p. 101.
[14] Pynchon, p. 583.
[15] *Ibid.*, p. 412.
[16] *Ibid.*, p. 413.
[17] Prigogine and Stengers, p. 54–5.
[18] Pynchon, pp. 318, 319.
[19] *Ibid.*, p. 324.
[20] *Ibid.*, pp. 326–7, 521.
[21] *Ibid.*, pp. 520, 524–5.
[22] *Ibid.*, pp. 203, 139.
[23] *Ibid.*, p. 74.
[24] Cooper, Peter L., *Signs and Symptoms: Thomas Pynchon and the Contemporary World* (University of California Press: Berkeley, 1983), p. 130.
[25] Prigogine and Stengers, p. 93–94.
[26] *Ibid.*, p. 12.
[27] Slade, Joseph, W., *Thomas Pynchon* (Warner Paperback Library: New York, 1974), p. 136, quoted in Marquez, p. 102.
[28] Marquez, p. 103.

[29] Pynchon, p. 638.
[30] Prigogine and Stengers, p. 95.
[31] Pynchon, p. 509.
[32] Prigogine and Stengers, p. 298.
[33] Hayles, N. Katherine, *Chaos and Order: Complex Dynamics in Literature and Science*, p. 27.
[34] Pynchon, p. 322.

RAYMOND J. WILSON III

METAPHORIC AND METONYMIC ALLEGORY: RICOEUR, JAKOBSON, AND THE POETRY OF W. B. YEATS

Elsewhere, I have pointed out that Paul Ricoeur denies symbolic effect to language situations where the original sense of an expression is destroyed in the process of interpretation, calling such discourse "allegory", the mere "rhetorical" and "didactic" procedure in which the literal meaning is "eliminated once it has done its job". The point emerged that Ricoeur needs two categories – one in which the initial sense is destroyed (Ricoeur's "allegory"), and the other in which the original meaning is retained and added to the second meaning (Ricoeur's "symbol"). If we are to have a dual paradigm, I argued that it is more helpful to subject symbolism to Roman Jakobson's paradigm, establishing two forms of symbolism, each signaled by contextual cues: the metaphoric symbol, in which the literal expression is discarded, and the metonymic symbol in which no such discarding occurs (Wilson 151). I concluded that under Jakobson's paradigm, Ricoeur's "allegory" is metaphorical symbolism and Ricoeur's "symbolism" is metonymic symbolism.

However, simply using the term "allegory" as an alternative to the term metaphoric symbolism encounters three problems. First, allegory derives meaning by relating the stated, overt pattern in the work to a known thought pattern, making allegory both more extensive than an individual symbol and also adding a dimension of familiarity that would appear to be the opposite of the "semantic shock" which characterizes metaphor in Ricoeur's system. Second, in Dante's letter to Can Grande, Dante describes four simultaneously operating meanings in allegory, and the literal meaning is not necessarily rejected as would happen in a metaphorical symbol (Musa 37). Third, we need the concept of allegory to explain aspects of the literal level; we want a concept of allegory which explains not only how the text provokes an allegorical meaning but how the allegorical insight re-illuminates the literal level. [1] William Butler Yeats's poems provide examples of the figures (metaphor and metonymy), as well as of the symbols (metaphoric symbolism and metonymic symbolism) with which we can develop a concept of

M. Kronegger and A-T. Tymieniecka (eds.), Analecta Husserliana XLII, pp. 219–227.

metaphoric allegory and metonymic allegory that will answer these questions.

Let us begin with the general principle. As Terence Hawkes interprets it, "Jakobson felt able to propose that human language in fact operates in terms of two fundamental dimensions whose characteristics are crystallized in the rhetorical devises of metaphor and metonymy" (Hawkes 78). If, as Hawkes points out, the metaphoric/metonymic split is basic to all language, if the distinction between them "reflects the fundamental dimensions of language itself", then we can hope that allegory as well as symbol can be explained by looking for the split (79).

Relating metaphor to similarity, Jakobson contrasts metaphor (something standing for another thing to which it is similar) to metonymy which Jakobson related to contiguity (something standing for another thing which is normally associated with it). Jakobson sees the writer as metaphorically selecting among several similar alternatives: "selection between alternatives implies the possibility of substituting one for another", implying that the unchosen alternatives thereby become available for metaphoric replacement, and thus, "selection and substitution are two faces of the same operation" (Jakobson and Halle 60).

Jakobson's metaphoric (replacement) principle applies to all of language which draws too large a circle to closely contain metaphor as a figure. To make the circle smaller, we must add the idea of difference to that of replacement. Writing of figurative metaphor, Paul Ricoeur notes that "metaphor is the result of the tension between two terms in a metaphorical utterance". In figurative metaphor, similarity competes with difference, as in Ricoeur's examples of "blue angelus" and "mantle of sorrow". Only "conflict" between two interpretations sustains the metaphor. Ricoeur saw "absurdity" as the strategy of the "metaphoric utterance", an absurdity that is exposed when the reader attempts a literal interpretation. "The angelus is not blue, if blue is a color" says Ricoeur; "sorrow is not a mantle, if the mantle is a garment made of cloth". The literal interpretation "self-destructs in a significant contradiction" (*Interpretation Theory* 50).

In preparation for the concept of metaphorical allegory, we need to illustrate figurative metaphor. If one says, "A cap is a hat", one does not make a metaphor, because not every similarity or identity is figuratively metaphoric. As the term "metaphor" has traditionally been used, only the equating – or close comparison – of literally unlike things creates metaphors. For example, in "Sailing to Byzantium", when William Butler Yeats says, "An aged man is but a paltry thing,/ A tattered coat upon a stick", we assume that the reader's mind intuitively and instantly rejects the literal reading since old men and coats literally are so different that the literal reading becomes impossible. "It is this process of self-destruction or transformation which imposes a sort of twist on the words, an extension of meaning thanks to which we can make sense where a literal interpretation would be literally nonsensical" (*Interpretation Theory* 50). Ricoeur points out that metaphor is not a word but a full sentence with definite predication, and that the predication flouts the "logical distance between two far flung semantic fields . . . creating a semantic shock which, in turn, sparks the meaning of the metaphor" ("Imagination" 7). After rejecting the literal reading, the reader, guided by what Ricoeur calls the strategy of "absurdity", finds an unexpected overlap in meaning, such as the worn-out ungainliness of both man and tattered coat (50).

Having employed Ricoeur's tension concept to identify the area covered by metaphor, within the circle of Jakobson's metaphoric principle, we can establish a metaphoric symbol to be like a metaphor except that the author writes only the vehicle and implies the predication which connects the vehicle to the (much different) tenor. The author implies by context – both textual and extra-textual. In Yeats's poem "The Second Coming", the narrator describes a "shape with lion body and the head of a man"; readers say the following: the coming historical era is a beast. We interpret the beast as a symbol because the literal interpretation makes no sense. Surrounding lines provide a context that signal the reader to discard the literal beast when the reader has understood that the beast stands for the fearful qualities of the age that Yeats sees as soon to commence. How, for example, could "twenty centuries of stony sleep" literally be "vexed to nightmare by a rocking cradle"? Abolition of the literal identifies the figure as metaphoric symbolism.

To clarify the relationship between metaphor and metaphoric symbol, suppose that, in "Sailing to Byzantium", Yeats had mentioned only the tattered coat on the stick and had not explicitly stated that an old man is a coat. What might be examples of contextual signals to the reader to decide that some of the characteristics of the coat overlap those of the old man? Consider that the poem opens with the following statement: "That is no country for old men. The young/ In one another's arms

. . . neglect/ Monuments of unaging intellect". From such lines, the reader might be cued to understand that no literal coat exists; he or she would then discard that meaning, recognizing that Yeats evokes it only to enhance our understanding of the man.

In moving from metaphoric symbolism to metaphoric allegory, we need the concept of a well-understood matrix into which the reader can plug the literal text, in order to derive the allegorical meaning. In *Symbolism and Interpretation*, Tzvetan Todorov shows how resolving contradiction between Christian doctrine and the literal text of the Bible motivated allegorical readings, for example readings of the Biblical reference to "Abraham's two wives". Early Christian interpreters used allegorical interpretations to avoid explaining why Abraham had two wives when other parts of the Bible clearly forbade polygamy by a metaphoric allegorical interpretation. "If," says Todorov, "'wives' were used to mean *weaknesses*", it would not be a case of forgetting the existence of the wives, but of "abolishing the initial statement: nothing would have been said about Abraham's (literal) wives (44)". In Todorov's construction, the discrepancy becomes the motive for the allegorical reading: a well-known meaning in an established structure (Christian doctrine) holds that God commands men to have only one wife, yet we read literally of a holy man who had more than one wife. The association of allegory with a well-known thought pattern is, thus, compatible with a metaphoric process of denying the literal meaning. This analysis demands the existence of a metaphorical allegory which destroys the literal meaning. A metonymic allegory would, then, be an allegory in which the literal level remained. Understanding metonymic allegory depends on grasping the essence of metonymic symbolism.

To define metonymic symbolism, we return to Jakobson's basic principles. Jakobson's metonymic principle draws too large a circle to neatly contain figurative metonymy, just as his metaphoric principle is too large for metaphor. To draw a smaller circle, within Jakobson's idea of contiguity, we can apply David Lodge's concept that figurative metonymy adds the notion of unusual choice to the contiguity of the metonymic principle. "Metonymy and synecdoche are produced by deleting one or more items from a natural combination, but not the items it would be most natural to omit (76)". For example, "Keels crossed the deep" contains metonymic remainders from the "notional sentence, *The keels of the ships crossed the deep sea*". If we wished to say, "The keels plowed the deep", *plowed* could only enter by metaphorically replacing *crossed*;

but the sentence contains redundancy, so that, under Jakobson's broad metonymic (contiguity) principle, some of its elements would become candidates for elimination – most likely *keels of* and *deep*, leaving the following: "The ships crossed the sea". The maker of a figurative metonymic sentence, in contrast, decides to keep these and eliminate *ships* and *sea*, instead, leaving "Keels crossed the deep". Just as is the metaphoric principle narrowed to figurative metaphor by adding "impertinent predication" to the guiding principle of similarity, so figurative metonymy is delineated by adding "unusual deletion" to the guiding principle of contiguity.

How does such a metonymic sequence become symbolic? This is more complex. An inadequate explanation – parallel to that of metaphoric symbolism – might be that the writer names only the part and does not name the whole. The non-naming explanation is inadequate because one never names the whole, even in simple metonymy: for example, in "Pardon old Fathers", Yeats asks his ancestors to forgive him for having, at the age of forty-nine "no child" – to forgive him for having "nothing but a book . . . to prove your blood and mine". Here, the book *does* exist; but, in addition, "book" stands for an entire lifetime of writings, contained in numerous books, magazines, and other media; a part (one book) stands for the whole (a career's worth of writing). However, even in this simple metonymy, the writing beyond that one book goes unmentioned. I would like to suggest that the concept of an "identifying association" will provide a more useful explanation of the process in metonymic symbolism.

As I previously have pointed out, in simple metonymy, images must be immediately "identifying" to be metonymically useful; not every part will automatically imply the whole (Wilson 155). For example, when a person writes that "keels crossed the sea", the author expects readers to know that the keels are present although "keels" also stand for "ships", part for whole. One would not, however, have the same expectation of, "The ropes crossed the sea" even though ropes are also parts of wooden sailing ships.

An author makes a metonymic symbol by placing an element in a transforming context; the writer takes an associated element, such as a part, that is *not* inherently identifying and makes it into an identifying element. This process retains the original meaning while adding a second meaning that is reached by the principle of association rather than a principle of absurdity. In Yeats's "The Wild Swans at Coole", no absurdity immediately forces us to discard pond and swans as we did Yeats's nightmare shape in "The Second Coming", and we apparently need not insist that the moment when the narrator stops to contemplate the swans is a symbol of lost youth and passion. We do not have simple metonymy since the moment is not in itself an identifying part of the sequence: "pond" is not automatically associated with a life pattern of "youth and passion", the way "keel" is associated with "ship".

In Yeats's "The Wild Swans at Coole", something leads us to transform the moment described into an identifying part of a larger life pattern. As in metaphoric symbolism, context provides the cue or cues for symbolic interpretation in metonymic symbolism. Perhaps we are cued by the speaker's words that having "looked upon those brilliant creatures", his "heart is sore". The speaker laments that "All's changed" since he trod this shore the first time with a "lighter tread"; perhaps this lament highlights how the swans remain "Unwearied still, lover by lover" while he has lost all intensity of passion. These contextual statements cue us to retain the literal event and place it within the sequence that it implies. If the narrator did literally "awake some day/ To find that they have flown away", he would literally feel sorrowful on observing the empty pond. The feared reaction is part of a whole, for which it can stand: the sequence of his threatened life pattern, structured by a desire to love and to be loved. The poem, which *really is* about a pause for a moment by the pond, portrays a tiny moment of a life so structured. When we feel that this moment stands for the entire life, we have decided that the moment is a metonymic symbol.

Removing the requirement for compared elements to be literally different, and thus the tension between similarity and difference, we envision the reader of metonymic symbolism moving through nesting levels of context, from a moment in a life to the entire life. We are guided by the principle of association, often the relationship of part to whole. This contrasts with the strategy of absurdity in metaphoric symbolism in which we are seeking similarity within literal difference.[2]

For readers of Yeats, his entire system acquires a familiarity that takes the place of an established meaning-pattern such as the Christian doctrine that the Church Fathers called upon to educe an allegorical reading of Abraham's two wives. In Yeats, "swans" always alert readers familiar with his poetry to "passion". The example demonstrates that the connection to the established meaning-pattern need not require the destruction of the literal meaning. Thus, a metonymic allegory would be consistent with Dante's system, as explained to Can Grande, in which the allegorical meaning coexists with the literal.

The interaction between metaphoric allegory and the metonymic variety allows the allegorical insight to re-illuminate the literal meaning, as is nicely illustrated by a few lines from Yeats's "Cuchulain's Fight With the Sea". In this poem, the hero broods after having been tricked into slaying his own son; his host asks Druids to "Chaunt in his ear delusions magical/ That he might fight the horses of the sea", rather than slay his friends in his uncontrollable grief and rage. The Druids do so, and after three days

> Cuchulain stirred,
> Stared on the horses of the sea, and heard
> The cars of battle and his own name cried;
> And fought with the invulnerable tide.

Both kinds of symbolism operate: (1) the horses of the sea stand for grief (metaphoric because, clearly, no actual horses exist), and (2) the tide stands for grief (metonymic because the tide continues to exist even after we realize what it signals). We come to see the vast horde of hallucinated enemies that the hero fights as standing for his grief. However, once we reach this understanding (grief equals enemy soldiers), we discard the horde, knowing that they are neither really present nor are they part of Yeats's point. This discarding identifies the connection as metaphoric symbolism. In the next line, the tide also stands for Cuchulain's emotional state: grief comes in, unstoppable like the tide, presenting the man on the beach the task of making sure he does not drown; yet grief, like the tide, will flow out again, equally independent of the will of the struggling man. We gain this new meaning (grief equals tide) without relinquishing our mental picture of the anguished warrior, standing knee-deep in the breakers, slashing at the incoming tide, a mental picture which is a powerful and essential feature of Yeats's poem.[3] This retaining of the literal image of the tidal flow while attaining the *additional* meaning of grief demonstrates how metonymic allegory makes the literal level more meaningful rather than less.

We are now in a position to assess the two sequences: metaphor/ metaphoric-symbol/metaphoric-allegory and metonymy/metonymic-symbol/metonymic-allegory. The replacement action of metaphoric symbolism fits with its requirement for sharp contrast between the overt vehicle and the implied tenor, Ricoeur's "self-destruction" of the literal interpretation; this operates as allegory when the motive for constructing the allegorical meaning is an inconsistency that exists between the literal meaning and a well-known meaning-pattern. But suppose a writer – the author of a realistic novel, for example – does not want to obliterate his or her literal meaning? By maintaining an original proposition which suggests *additional* meanings, metonymic symbolism presents the advantage of multivalent-meaning. This becomes allegorical when for example in Melville's *Billy Budd*, the main character's actions take on added significance when seen as superimposed on a well-known pattern, in this case of Christ.

Applying Jakobson's metaphoric-metonymic system to allegory thus satisfies the three problems presented by Ricoeur's designating as "mere allegory" those aspects of language that do not fit his definition of symbol. First, a metaphoric allegory is consistent with allegory's association with a well-known pattern of thought. Second, a metonymic allegory is consistent with Dante's concept that allegory presents several simultaneous meanings that do not replace the literal meaning. Third, a metonymic allegory, interacting with the metaphoric variety, allows the allegorical insight to re-illuminate the text, enriching all elements of the literal meaning.

Loras College
Dubuque, Iowa

NOTES

[1] I am indebted to the participants of the sixteenth annual convention of the International Society of Phenomenology and Literature for these three critiques.

[2] To see the metaphoric/metonymic concept in context, we relate metonymic symbolism to what, in *Symbolism and Interpretation*, Tzvetan Todorov calls "*propositional symbolism*", in which case metaphoric symbolism would be similar to what Todorov calls "*lexical symbolism*". A Biblical passage that states "animals will be rewarded" serves as Todorov's starting point; commentators using what Todorov calls propositional symbolism took this to mean that men will be rewarded while those using lexical symbolism decided it indicated that only the meek will be rewarded. In the propositional symbolic interpretation, Todorov finds "three propositions": "(1) animals will be rewarded; (2) men are like animals (only better); (3) men will be rewarded too". In maintaining the original meaning, propositional symbolism is essentially metonymic. In the lexical symbolic reading, Todorov finds only "two propositions": "(1) some men are meek like animals; (2) these men will be rewarded" (40). The missing proposition

in lexical symbolism is the original statement in the text, indicating that this interpretation *replaces* the word "animal" with the expression "men who are meek like animals". The lexical symbolic interpretation is thus metaphoric in allowing readers to dismiss the literal statement, freeing them from doctrinal embarrassment of heavenly rewards for animals.

[3] For Todorov the creation of the second meaning, while retaining the first, is the crucial element that separates this form of symbolism from non-symbolism. Using as an example, "the lances came into the room", meaning "the lancers came into the room", Todorov explains that a simple figure gives "two descriptions of the *same fact* (the entrance of the lancers)". On the other hand, symbolism gives "the description of a *second fact*: from the fact that animals will be rewarded we conclude that men will be rewarded too" (40).

REFERENCES

Hawkes, Terence, *Metaphor*, The Critical Idiom (London: Methuen, 1972).

Jakobson, Roman, "Linguistics and Poetics," in *Style in Language*. Ed. Thomas Sebeok (Cambridge: MIT Press, 1960), pp. 350–77.

Jakobson, Roman and Morris Halle, *Fundamentals of Language* (The Hague: Mouton, 1956).

Lodge, David, *The Modes of Modern Writing: Metaphor, Metonomy, and the Typology of Modern Literature* (Chicago: University of Chicago Press, 1977).

Musa, Mark, Ed. "Dante's Letter to Con Grande" trans. Nancy Howe, in *Essays on Dante* (Bloomington: Indiana University Press, 1964).

Ricoeur, Paul, *The Symbolism of Evil* (1967). Trans. Emerson Buchanon (Boston: Beacon, 1969).

Ricoeur, Paul, *Interpretation Theory: Discourse and the Surplus of Meaning* (Fort Worth: Texas Christian University Press, 1976).

Ricoeur, Paul, *The Rule of Metaphor: Multidisciplinary Studies of the Creation of Meaning in Language* (1975). Trans. Robert Czerny with Kathleen McLaughlin and John Costello (Toronto: University of Toronto Press, 1977).

Ricoeur, Paul, "Imagination in Discourse and Action," in *The Human Being in Action: The Irreducible Element in Man II*. Ed. Anna-Teresa Tymieniecka. *Analecta Husserliana* (Dordrecht, Holland: D. Reidel, 1978), pp. 3–22.

Todorov, Tzvetan, *Symbolism and Interpretation* (Ithaca: Cornell, 1982). French publication, 1978.

Wilson, Raymond J. III, "Ricoeur's 'Allegory' and Jakobson's Metaphoric/Metonymic Principles," in *Phenomenological Inquiry* **15** (October 1991): pp. 151–161.

STARLA STENSAAS

THE TRANSFORMED AND TRANSFORMING IMAGE IN THE SHIFT FROM PRINT TO DIGITAL CULTURE

The images in and of a culture reflect our cultural understandings of ourselves. Images in any culture are constantly changing and as they change, they transform the culture. Likewise, as a culture undergoes change, the images of the culture are transformed. In this way, image and how we read images are both the trace element of and a visual wake following the shifts in our collective understandings. Image operates both as an integral part of an organic, changing energy system that generates momentum towards a cultural shift, and as a footprint that references visual metaphors in describing a cultural shift that is taking – or has taken – place. To understand the transformed and transforming nature of images in culture, we must first understand the components of the image: the compositional or visual language, the cultural context of the image, and the medium in which the image appears.

The image is known first as visual language (whether acknowledged or not by the viewer) complete with a grammar and intertextual relationships of the parts to the whole. Just as readers come to a text, we as viewers come to the image with a range of literacies for "reading" the visual language which is the foundation for knowing the image. Although we may settle for an easy "reading", leaving the complexities of visual relationships behind, we still know the image first through the elements and principles of design – that is, visual language – just as we as readers of a text understand words in the context of letters, sentences in the context of subject and verb, and character development in the context of plot.

The "grammar" of visual language is made up of the elements and principles of design. The elements – line, shape, texture, color or value, and depth – are the parts that make up the whole. The principles of design – unity and variety, balance, rhythm, focal point, proportion and scale – are the relationships among the design elements, to one another and to the whole. In two-dimensional representational images, design elements are the foundation that sustains the illusion of the representation, the foundation that creates the symbol for the subject matter, allowing the viewer to recognize a face, a bowl, or a tree, for example.

M. Kronegger and A-T. Tymieniecka (eds.), Analecta Husserliana XLII, pp. 229–247.

Without the elements of design, such as the range of values in a black and white photograph that creates the illusion of roundness, or a line that separates the figure from ground in a lithograph, the illusion of representation would not be possible. All representational images, before they are the illusion of the things that they represent, are combinations of the basic design elements of line, shape, texture, color or value, and/or depth.

It is the design principles that create a context in which the design elements interact with one another. The possible combinations of design relationships that can appear in an image are as many and as complex as the human imagination is open to possibility, but all images contain – more or less successfully – a range of design principles. Unity and variety, for example, can be found in any image. A sense of unity must be sustained in order to create a sense of coherence in the image – possibly through repetition in linear direction, thickness, length, or gesture, for example, or the use of a reoccurring color or motif. In order to sustain viewer interest, an image must also contain a degree of variety, or difference – possibly through the use of contrasting lines or kinds of shape, through the inclusion of a variety of values or intensities of color, or through shifts in texture, for example. The quality of an image as a whole is affected by the relationships of the design elements to one another and to the whole, that is, to the use of design principles in a composition.

In non-representational images, it is often the visual relationships between design elements themselves which are the "subject matter" of the composition. In the work of Abstract Expressionists (1945–60), for example, it is the interaction of line, shape, color and texture that conveys the emotive qualities, or "subject", of the work, which has been described as vibrant with energy, overwhelmingly dynamic – and even violent – in presence, bold, passionate and fierce.[1] Likewise, it is the element of color in relationship to itself that is the "subject" of Color Field painting, such as the work of Mark Rothko in the late 1940's to early 60's. Although the choice of line weight and placement in a Mondrian places a great deal of emphasis on the element itself (the line), it is the relationship between the line and the geometric shape of color and the implications of that relationship on the balance, rhythm, and variety in the work that makes the line such a striking element.

In terms of design elements and principles alone, the *Mona Lisa*[2] is known as a compositional whole, whether she appears as Duchamp's

L.H.O.O.Q.[3] with mustache or as part of an advertisement for Prince spaghetti sauce. The visual language changes subtly as an image moves from its original context – in this case oil on canvas – to the alternate context: for example, scale and surface texture changes when the *Mona Lisa* appears in a magazine advertisement. Nonetheless, the image in this example remains fairly constant as a compositional whole. This same subtle compositional change occurs when, for example, O'Keeffe or Dali compositions are seen as four-color printed images on coated stock in art books rather than the original oil on canvas. These two examples demonstrate an opposition which is germane to this particular compositional concern: O'Keeffe's work becomes slick and almost expansive in printed plates whereas her canvases are texturally intimate; and some of the intensity and slickness of the Dali canvases expressed in rich luminous paint in small meticulous renderings are lost in the printed plate. But despite this subtle compositional change as the image moves out of its original context, in all of these examples the basic visual language remains intact. In either context the line placed by the artist's hand onto the canvas appears in the same relationship to the other lines and shapes in the picture plane; elements carry the same visual weight and are engaged in the same complexity of visual relationships.

Yet, we do not see the image in a visual paradigm alone, but in a complexity of cultural contexts. Because the collective we, as a culture, come to an image with an awareness of the historical construct of a particular time period or the cultural construct of what a particular image, such as the *Mona Lisa*, has come to mean over time, an image is changed through placement in a new cultural context. That is, the visual language of the image may remain constant, but our sense of what the image means changes with the change in cultural placement. This is especially true when the intention of the new placement is to reference something besides the visual language of the original, such as the use of the *Mona Lisa* to sell pasta sauce.[4] Even the viewer uninitiated to the specific historical traditions surrounding the *Mona Lisa*, recognizes the *Mona Lisa* as a cultural icon and finds her image, having gained a hefty fifty or more pounds to represent "chunky" pasta sauce, incongruous and humorous. Part of the humor is the comparison of the new "chunky" version with the "original" – the *Mona Lisa* as we know her, intact and presented coupled with the original version of the pasta sauce. The original pasta sauce *Mona Lisa* may look the same, have the same visual grammar, as the original oil on canvas *Mona Lisa*, but what she means, represents – even becomes – changes significantly in the new context.

Art historian Arlene Raven points to the mystique surrounding the *Mona Lisa* smile as a cultural one. She and other art historians note that what an image comes to mean over time is a reflection of the cultural assumptions of the viewers who examine the image. Raven suggests that the Mona Lisa smile is an "archaic smile" that connotes in its passivity and muteness a cultural assumption regarding the "essence of womanhood." She writes:

In the arts, as elsewhere, female facades were not routinely examined for what might lurk beneath, nor even reflected upon as authentic human appearances. Consider when Marcel Duchamp added a mustache to an image of Leonardo's *Mona Lisa* in 1919. His deliberate affront to female beauty is not usually seen as central to his gesture. The younger artist is perceived, rather, as actively responding with primal jealousy to the great art of the past. Female beauty, he also demonstrated, is a commodity that depends on absolute passivity and the absence of character or human persona.[5]

The significance of Duchamp's gesture in creating *L.H.O.O.Q.* has less to do with his intention or the compositional integrity of the *Mona Lisa* than it does with the cultural implications of and reception to his new placement of the *Mona Lisa* in a context outside of the cultural mystique that then surrounded her. Similarly, Manet may have "regarded the nude figure as a study in painted form, an exercise to display his newfound insights into light and shape,"[6] but when he placed a nude woman in the presence of two clothed men in *Luncheon on the Grass* (*Déjeuner sur L'Herbe*)[7] in 1863, the nude outside of the classical guise of goddess, harem slave, or allegorical figure created a scandal. Had Manet's contemporaries known that he had "borrowed" the compositional grammar of Raphael's *The Judgement of Paris*, 1520, art historian Janson supposes, *Luncheon on the Grass* might have seemed "less disreputable . . . since the hallowed shade of Raphael could be seen to hover nearby as a sort of chaperone."[8] Regardless of Manet's compositional intentions, the meaning embedded in Raphael's original visual grammar was changed significantly by Manet's introduction of clothes on the male figures emphasizing the contrast to the nude female figure – in this instance because of the cultural assumptions at that time with regard to the appropriate use of the nude in a fine arts context. Maintaining the basic compositional integrity of an image's visual language does not maintain the integrity of the image as a whole; the cultural context and placement of the image (particularly in relation to the intention of the new placement), significantly change the way we see the original image in a new placement.

Likewise, cultural context can be an underlying framework from which an artist develops a new image. Referencing the visual language of a familiar image significantly affects the way in which we view the new image, indeed, informs our understanding of the new image. For example, John De Andrea's polyvinyl, polychromed and oil sculpture, *Manet: Déjeuner sur L'Herbe*,[9] 1982, imitates the figures of Manet's *Luncheon on the Grass* by placing the clothed male figures of the artist and his assistant in life-size relief next to a nude female model on a plaster splattered platform. Confronting these life-sized figures in polyvinyl – a process that uses the human as a mold, thereby creating a lifelike illusion of the figure – in the art gallery or museum is as shocking to the contemporary viewer as confronting Manet's canvas was to his public. One does not expect to see plaster-encrusted workmen resting in a formal museum environment; nor is one comfortable with the nonchalance of the male figures on an apparent break from some physical labor seated next to a nude female figure. De Andrea has referenced the cultural context of Manet's *Luncheon on the Grass* by imitating or more accurately, "quoting" – the placement of Manet's figures through visual language, embedding the new work with a particular and specific meaning in its cultural reference to the Manet.

Similarly, Jean Fraser in her series *Celestial Bodies*[10] uses Manet's *Luncheon on the Grass* to inform one of a series of photographs, with text, that explore "the socially constructed nature of sexuality"[11] by referencing Catholic nuns as a visual and conceptual metaphor. Six photographs of two nuns and sometimes the Manet-like nude female appear in the series, coupled with type. One of the photographs imitates Manet's figures in *Luncheon on the Grass* by placing a nude woman holding a glass of wine next to two clothed nuns. Type above and below the image reads "blasphemy" and "communion." In this instance, Fraser has used the Manet as a startling cultural counterpoint in which issues of gender and sexual orientation stand in stark relief to the heterosexual iconography of clothed men in relation to the nude female. In the introduction to the book, *Stolen Glances: Lesbians Take Photographs*, in which the series appears, the authors write: "Our imagination was caught by the inventiveness of lesbian photographers who had 'stolen' and inverted the meanings of mainstream, heterosexual imagery."[12] Fraser's "invertedness" references a heterosexual icon in a culture that

Adrienne Rich and others have described as a compulsorily heterosexist,[13] and points to the socially constructed nature of heterosexuality. It is this social construction portrayed in Manet's *Luncheon on the Grass*, then, that becomes the cultural context that is referenced or "quoted" in Fraser's work.

In both of these examples not only does the Manet inform the work which references it, but the way in which we view the original work is also changed by the new placement in varied cultural contexts. De Andrea's *Manet: Déjeuner sur L'Herbe* re-establishes for the contemporary viewer the shock – especially in a fine arts context – of seeing a nude woman seated next to nonchalant clothed men. Because of our continual exposure to photographic images, the contemporary viewer does not necessarily experience the same sense of shock as Manet's contemporaries did in viewing the two-dimensional, oil on canvas representation of the figures. Contemporary culture has inundated viewers with photographic images and the nude, or nearly nude, female is frequently seen in a wide variety of contexts from selling car parts to perfume, from engaging in photographically explicit sexual fantasies to television mini-series. De Andrea has broken through the veil of our contemporary cultural dismissal of such two-dimensional representational images as "other" or "not real," learned through our repeated exposure to them, and has confronted the viewer with an illusion of reality that reinterprets Manet's *Luncheon on the Grass* in a context accessible to contemporary viewers. The original cultural context of the two clothed men seated next to the nude woman is evoked and made accessible to the contemporary viewer through the use of lifesized and illusionary lifelike figures in a placement that is familiar and yet out of context in the museum or gallery setting. In this way, De Andrea's image becomes a footprint that references the changes in how we have seen and now see this particular arrangement of figures. He has re-established the vital cultural context for the contemporary viewer to re-visit the Manet image, to see anew the original cultural context of that image.

Fraser, on the other hand, changes the ways in which we view the Manet not by revisiting, but by revisioning the cultural context of the image. Her reference to *Luncheon on the Grass* becomes part of an organic, changing energy system that generates momentum towards a cultural shift – in this case a shift towards understanding the socially constructed nature of sexuality, specifically in a compulsorily heterosexist culture. After viewing Fraser's "inverted" use of Manet's placement of clothed men in relation to the nude women, the objectification of the woman and the privilege of the dressed men as a heterosexual icon becomes clear; it is the inversion and revision of the heterosexual icon that gives us pause to consider its meaning *as* a heterosexual icon. Of course Fraser's work as a lesbian photographer goes beyond a cultural critique of the heterosexual construction of Manet's compositional arrangement, but it is her reference to Manet's work in this regard that opens the door to an alternative way of seeing the Manet. Manet's *Luncheon on the Grass* is transformed through Fraser's revisioning of the image, by the cultural context she uses as the underlying framework for her image.

What sits between the cultural context and the visual language of the image is the medium in which the image appears. It is not that the medium *is* the message as Marshall McLuhan[14] has suggested: rather the medium creates both a cultural context and a medium-particular visual language, contributing significantly to the meaning we "read" from an image. A line drawn with a pencil is not the same, compositionally speaking, as a line etched with acid onto an intaglio plate, nor a line laid down with a full brush of aqueous pigment. Not only are the gesture, characteristic, and quality of the line as part of the visual grammar of the work changed (that is, one line is fat, the other delicate and fragile, another is full of variations of thick to thin), but the cultural context of the line changes.

The historical position of painting, for example, lends an aura of importance to the medium that is not found in drawing or printmaking. Historically, painting developed a singular importance and has stood as the primary footprint for historians in describing art history. Drawing, on the other hand, was historically considered "merely" a study for the more important painting, and imparts its secondary positioning through that cultural context, referenced visually by the mere fact of the mark made by pencil, pen, chalk or crayon across the page. Prints, due to their multiplicity and small scale, were more closely tied to text in the book form, removing them historically from the domain of art that hung on the church or castle wall and likewise separating them in importance from painting or sculpture.

Within the historical and cultural distinctions between media in the fine arts come lengthy constructions of similarity and differences that circle back to visual language as a source for describing the differences: painters traditionally distinguish shape by shifts in value, texture, or color, creating shape through the use of edges; printmakers traditionally use a linear description, an actual line, to designate the edge of shape. In recent decades, prints and photographs have pushed the technical boundaries of the once limited and smallar scale which has been used to designate these media as less important than the "larger" painting. Each medium has a visual language of its own and those distinctions in visual language both identify the medium with, and create, particular historical and cultural contexts.

The interesting questions that arise in looking at the medium in which the image appears, however, are not ones of value judgement – this medium is better or more important than that one – nor is the issue of medium limited to the narrow confines of the image in a fine arts context. Medium in a larger sense designates the boundaries of our collective understandings – our epistemology – in the ways in which it is integral to both how we "read" the cultural context and visual language of image and text, and how that constructed "reading" defines our thought idiom. Just as the medium in a fine arts context creates a medium-specific visual grammar and a cultural context from which the artist is able to shape an image from his or her own inner vision or knowing, so too does medium in the larger sense define what kinds of "pictures" we collectively create to understand or know the world. When Käthe Kollwitz, whose genius was her draftsmanship – her ability to use line and shading to convey deep human emotion, portrayed the figure in extreme anguish in *Fettered Man*,[15] 1927, it was the bold black of brushed ink that gave her the "grammar" with which to express such anguish. In contrast, it was pencil in *Karl Liebknecht on his Deathbed*[16] in 1919 that gave her the subtle grays and delicate lines from which the figure emerges nearly lifeless and barely visible. Medium profoundly affects the kind of image that can be made and, perhaps, even the kinds of images that can be imagined.

The most striking example of the effect of medium in the large, cultural sense is found in the shift that occurred in the move from the oral-aural communication paradigm to the literate, print culture communication paradigm. Walter Ong in *Orality and Literacy: The Technologizing of the World* writes:

Fully literate persons can only with great difficulty imagine what a primary oral culture is like, that is, a culture with no knowledge whatsoever of writing or even of the possibility of writing. Try to imagine a culture where no one has ever "looked up" anything. In a primary oral culture, the expression "to look up something" is an empty phrase: it would have no conceivable meaning. Without writing, words as such have no visual presence, even when the objects they represent are visual. They are sounds. You might "call" them back – "recall" them. But there is nowhere to look for them.[17]

Difficult as it may be for a literate person to imagine what a primary oral culture is like, it is equally difficult to imagine what lies beyond literate, print culture. Michael Heim in *Electric Language: A Philosophical Study of Word Processing* considers the implications of a shift from literate, print culture to digital culture as it relates to the phenomenon of word processing. Word processing, he hypothesizes, changes the ways in which we write, and therefore think about the world, much the same as writing changed the ways in which primary oral cultures were able to think about their world. He discusses Eric Havelock's work, as well as others, as a philosophical basis for understanding how the relationship between orality and literacy might inform what lies beyond literacy. He writes:

The "communication technology" of poetry fosters an idiom of speaking and thinking in the culture as a whole. The "mode of consciousness", the vocabulary and syntax of oral culture, is not the same as that cultivated by a society which regards books as the paradigm of significant communication worth preserving . . . The thought idiom based solely on repetitive acoustic patterns and on sharp narrative images strongly reinforces an empathic, tribal emotional identification (Greek *mimesis*) with admired cultural heroes.[18]

Ong further outlines some of the characteristics of orally based thought and expression noting that "in an oral culture, restriction of words to sound determines not only modes of expression but also thought processes. You know what you can recall."[19] In an oral culture, sustained thought is tied to communication – an interlocutor is required. In addition, mnemonic patterns are central to solving the problem of retaining and retrieving thought. Rhythmic patterns are also useful in aiding recall, as are formulas. In other words, "in an oral culture, to think through something in non-formulaic, non-patterned, non-mnemonic terms, even if it were possible, would be a waste of time, for such thought, once worked through, could never be recovered with effectiveness, as it could be with the aid of writing. It would not be abiding knowledge but imply a passing thought, however complex."[20] In contrast, literate, print culture breaks away from an immediate emotional identification through the use of patterned, formulaic mnemonics and fashions "abstract concepts capable of subsuming immediate realities under more general ideas;" Heim writes, "Plato's protest against poetry was on behalf of thinking itself, in the philosophical sense, which requires a different paradigm than that offered by oral models."[21] Clearly, like the difference between the inked brush stroke in *Fettered Man* and the pencil mark in *Karl Liebknecht on his Deathbed*, the difference in the "picture" created by thinking in an primary oral culture and the "picture" created by thinking in a literate print culture is significant. Print culture, print as a medium, changed forever the processes by which we are enabled to consider and think about the world.

Many are now writing about a shift from print culture to digital culture. Michael Joyce, in "Notes towards an Unwritten Non-Linear Electronic Text, The Ends of Print Culture," considers the implications of the digital medium in relation to fiction, describing an environment for writing and reading fiction in which the interaction between writer and reader at the least empowers the reader, and at its best is reciprocal, with reader-as-writer. The digital environment which he describes, often referred to as hypermedia, is familiar to most who have encountered it as HyperCard, an Apple software product (although there are other hypertext products such as Storyspace, Supercard, Guide and ToolBook, for example).[22] In such an environment, the viewer/reader is involved with a "stack", a number of related computer screens, which may include text, sound, graphics, video, and/or animation. The "hyper" of hypermedia refers to "nonlinear methods of moving through a body of information. One idea or image gives rise to many associated ideas and images, which in turn give rise to others, rather than moving in a linear fashion, as in a traditional novel."[23] The viewer/reader, once empowered to interact with the computer hardware by grasping some understanding of the software, can browse through a stack according to his or her own inclinations, investigating associations between bits of information in a non-linear fashion.

Joyce suggests that the digital medium, even at its least reciprocal, pushes us towards a thought idiom that embodies multiplicity. "Already with remote control channel zapper in hand," he writes, "most of us can track multiple narratives, headline loops, and touchdown drives simultaneously across cable transmissions and stratified time . . . we can shift values everywhere, instantly, individually, as we will."[24] In the shift from print culture linearity towards a digital culture that embraces multiplicity, "picture" making as a poetic process is expanded and enlarged. At its most reciprocal, the digital medium doesn't merely define the kind of image the artist or author is able to shape and the boundaries within which the image is read; it also opens the shaping of the image itself to both the artist/author and the viewer/reader. In this context, the digital medium can be an environment in which the image or text no longer travels as a compositional whole changed by its placement in varying cultural contexts – rather, varying cultural contexts literally change the compositional whole. That is, the viewer/reader is able to alter the compositional whole of the image, and will do so within the parameters of his or her cultural contexts; the viewer/reader can literally embed the grammar of the work with a new cultural context.

Even at its least reciprocal, the digital medium appears to break down the dualism of either/or inherent in other media through a kind of simultaneity that reveals the possibility of many – for one example, through creating the potential of watching several narratives simultaneously by way of the television remote control. Imagine how watching *The Sound of Music* on one channel, cut in segments the length of time between commercial interruptions, might inform and change the viewing of Woody Allen's *Annie Hall* on another channel – a very different experience from the original cultural one of watching either film in a movie theater.

In print culture, we may describe the image or text as a unified whole shaped through the artist/author's act of knowing, subsuming the viewer/reader who must enters this unified whole; or we may describe the viewer/reader as consuming an image or text (created as a unified whole from the artist/author's act of knowing) and then subsuming it into their own understanding of the world. In either case, within the language of print culture, the creative process is described as linear, beginning with the making and ending with consumption, and a larger, print culture particular "picture" of the world is created in which consumption and subsumption are appropriate responses. Digital culture, on the other hand, offers the possibility of multiplicity, an endless succession of illuminated "readings" in which the readings themselves, rather than consuming and subsuming, remake the original, where viewing/reading becomes – within the context of the expression of the artist/author's inner vision or knowing – a creative act that describes the viewer/reader's inner vision or knowing as well.

In considering the issues germane to "reading" images in a digital medium, it is useful to look at the characteristics of digital visual language, noting the similarities and differences between it and the more familiar print culture visual language. Many HyperCard presentations have incorporated imagery that uses a pre-digital syntax – that is, visual language that references print culture media – such as John McDaid's hyperfiction *Uncle Buddy's Phantom Fun House*,[25] an electronic environment built from photographs, Tarot cards, souvenirs, and a sketchbook of drawings and text. In the fiction, the narrator has left this material on a computer hard drive and the reader comes into the fiction by way of an opening letter signed by the narrators' lawyers which details the facts regarding the disappearance of the narrator and how the fiction has come, ultimately, into the readers' hands. Many of McDaid's images, especially the photographic ones, have a digitally-appropriate syntax to them, being similar in their compositional grammar to television images. But other images, particularly those in the sketchbook, reference a pre-digital syntax by attempting to imitate the gesture of a pen or pencil moved across a sheet of paper. Even more disconcerting is text, as if it has been rolled out of a typewriter, on the opposite "page" from the "pencil" drawing in the "sketchbook". The illusion of page, pencil, and sketchbook, particularly with the incongruently placed "typewritten" text, lacks verisimilitude in the context of material left on a computer hard drive. Why is the visual language of print culture referenced within this digital fiction?

In fact it seems that we, including artists and computer hardware designers marketing to artists, are intent on forcing the visual language of other print culture media into this new one. Joyce notes a similar longing in relation to the book as artifact:

> The reader struggles against the electronic book. "But you can't read it in bed," she says, everyone's last ditch argument. Fully a year after Sony first showed Discman, a portable mini-CD the size of a Walkman, capable of holding 100,000 pages of text, a discussion on the Gutenberg computer network wanted to move the last ditch a little further. The smell of ink, one writer suggested; the crinkle of pages, suggests another.
>
> Meanwhile in far-off laboratories of the Military-Infotainment Complex – to advance upon Stuart Moultrop's phrase (Moulthrop, 1989b) – at Warner, Disney or IBApple and MicroLotus, some scientists work on synchronous smell-o-vision with real time simulated fragrance degradation shifting from fresh ink to old mold; while others build raised-text touch screens with laterally facing windows that look and turn like pages, crinkling and sighing as they turn. "But the dog can't eat it," someone protests, and – smiling, silently – the scientists go back to their laboratories, bags of silicone kibbles over their shoulders.
>
> What we whiff is not the smell of ink but the smell of loss . . . [26]

Artists, trying to force a pre-digital drawing syntax – the print culture methods of drawing – into this new medium, complained that the shape of the mouse is not the same as the shape of a pencil; the mouse doesn't allow for the freedom of expression that the pencil does. So the hardware designers created drawing styluses in the shape of pencils, but the artists complained that the drawing stylus is too limiting; it doesn't allow for the nuance of the hand's pressure to paper surface. The hardware designers invented a stylus that creates different levels of gray depending on the pressure of the stylus to tablet. But the artist still complains that drawing is unnatural with the eye trained to the left on a computer monitor while the hand draws to the right on a tablet. And when I look at a clumsy line that doesn't really look anything like a pencil line, on the vibrating white of the computer monitor HyperCard screen that doesn't look anything like paper, I, too, wonder why the narrator in *Uncle Buddy's Phantom Fun House* chose to leave an electronic "sketchbook" instead of a paper one. Or why drawing is so dogmatically privileged by artists as an absolutely fundamental language for visual expression, giving rise to an expectation of its primacy in the digital medium as well.

Perhaps drawing is privileged because we have all drawn as children – or, perhaps because it has been the primary language that artists have historically used to work out visual ideas, especially as studies for paintings. But drawing is no more a fundamental language in the digital medium than it is in photography, film, or television. So what does it mean when a digital novel references drawing and, in doing so, the visual language of print culture? Is the "sketchbook" in *Uncle Buddy's Phantom Fun House* transformed as a cultural icon by its new placement in a digital context? Or does it revisit print culture, like De Andrea's use of Manet's placement of the figures in *Luncheon on the Grass*, and become a footprint describing the shifting ground between print and digital culture, somehow making the landscape safer in the transition between the two? These are difficult questions in that they ask us to imagine what lies beyond print culture while still speaking in the grammar of print culture.

Other questions are, perhaps, more easily considered in that they suggest a comparison with a print culture counterpart. If a pre-digital syntax, for example, particularly in its reference to visual language from other media, is not the visual language of the digital environment, what does a digital visual language look like? What kind of visual language is being fashioned in this new medium and what particular historical and cultural contexts will be created and identified in using this new visual grammar? As with fine arts media, looking at the processes and technical constraints inherent in the medium can point to specific characteristics of that medium's visual language: charcoal, for example, because it has no binder is friable and needs only light pressure to mark a surface allowing for delicate lines or bold broad tones with a great degree of freedom to blend values and change or erase marks; on the other hand crayons, which contain a strong adhesive binder, create marks that are almost impossible to remove from the page, making the blending of tones more difficult and creating "sharp, rich lines . . . which like pen and ink, demands resolute decisions and handling."[27] A medium that allows for a great degree of freedom to alter the image offers the artist a much different potential for image-making than a medium that requires resolute decisions and handling. The difference between what is conveyed by the inked brush stroke in *Fettered Man* and the pencil mark in *Karl Liebknecht on his Deathbed* is tied to the processes and technical constraints of the medium being used.

Likewise we can suppose some characteristic visual language in the digital environment, in addition to the broad, sweeping cultural associations with multiplicity and simultaneity, that is based in the hardware and software constraints of the medium. Some of these characteristics overlap the visual language of television and, to a lesser extent, film due to similarities in the potential for movement as well as how an image is displayed on the computer monitor. Pixel resolution, the aspect ratio of horizontal to vertical pixel display, and the size of the monitor all impact significantly on the kind of visual language the viewer experiences looking at a digital image. In both a television image and an image projected on a computer monitor, for example, the viewer experiences the image primarily to the center of the screen or just off center. In contrast, film allows for a greater play of images in the background and to the edges of the screen, potentially creating a wider and more subtle visual vocabulary than the television/computer-generated image where background and peripheral images are often disregarded or tuned out.

This occurs because of the technical constraints of video versus film. The "manner in which the electronic formation of moving images differs from the photographic basis used in cinema" changes the visual grammar of the final image.[28] L. Bernard Happé in *Basic Motion Picture Technology* explains the technical differences between film, which is a photographic process, and television, which is an electronic one:

While the film frame contains a complete two-dimensional record of the whole of the scene before the camera at any one moment of time, television is concerned at any instant with the light and shade of only one point in that scene. In order to build up the whole picture every point in the scene must be examined, but this is done in an orderly fashion by scanning a series of horizontal lines which form a grid pattern called the 'raster'. The brightness of each point along the line is transmitted as a successive series of electronic signals and when the scanning of one line is complete the process is begun again at the start of the next horizontal line. When the whole series of lines comprising one full picture area has been completed, the scanning process commences at the first line once more. . . . As in the cinema, a successive series of picture images must be presented rapidly one after the other to give the sensation of motion, but unlike the cinema, there is no period between successive pictures when the screen is dark.[29]

Englander and Petzold in *Filming for Television* consider the implications of these technical differences between film and television on the composition of the television image:

Although there must be much in common between composition in films intended for cinema and those destined for television showing, particular techniques are applied in television work to compensate for the fact that the image is magnified to only 1/2000th part of the viewing area of a large cinema screen. Therefore, with television filming, as opposed to filming for the screen, when composing the picture you make a point of cutting out all the "air" in your compositions. . . .

You cannot effectively photograph wide open views with tiny figures seen in the distance because such detail is lost on a television screen.[30]

Clearly the process by which images are transmitted and the size of the screen on which they appear affect the kinds of images seen on television and have implications regarding the kinds of images we see and will be seeing on the computer monitor. That we see more to the center of the screen and with less subtlety on a television or computer monitor than when viewing a film image surely affects the kinds of messages the digital medium can and does convey. On the other hand, digital color on the computer monitor has a much wider range than television and that capability will certainly delineate between television's visual language and the developing computer-generated visual language, until, that is, high res and stereophonic television is combined with interactive home computing, eliminating any delineation between the two media at all.

There is also a cultural context that informs our viewing of television that carries over to the digital medium, affecting our expectation of what is appropriate digital visual language. David Antin in *Video: The Distinctive Feature of the Medium* writes:

The social conventions of television dictate a code of behavior according to which the transmitter must assume two apparently different roles in transmission. In one he must appear to address the viewer on the station's behalf as an entertainer; in the other on the sponsor's behalf as a salesman. The rules of the game, which are legally codified, prescribe a sharp demarcation between the roles, and the industry makes a great show of marking off the boundaries between its two types of performances – the programs and the commercials . . . Since the industry's most tangible presentation is metrically divisible time, the industry seems to mark the separation emphatically by assigning the two roles different time signatures. The commercial is built on a scale of the minute out of multiple 10-second units . . . The program is built on the scale of the hour out of truncated and hinged 15-minute units.[31]

Antin goes on to argue that television's metrically divided time has created a social convention which we have come to associate with video, creating an expectation of what the visual language of video ought to look like. "It may not be quite hip to say so without qualification," Antin writes, "but it is a commonplace to describe artists' videotapes as 'boring' or 'long,' even when one feels that this in no way invalidates or dishonors the tapes in question . . . This perceived quality of being boring or long has little to do with the actual length of the tapes."[32]

Television's division of time can be considered a formulary device in what Ong has called "secondary orality." "I would suggest," he writes, "that we speak of the orality of preliterate man as primary orality and of the orality of our electronic technologized culture as secondary orality. Secondary orality is founded on – though it departs from – the individualized introversion of the age of writing, print, and rationalism which intervened between it and primary orality and which remains as part of us."[33] Ong defines formulary devices as "any set or standardized verbal expression, 'set' and 'standardized' implying, of course, that the expression is used more than once and generally quite often."[34] Formulary devices in primary orality were used "chiefly as knowledge storage and retrieval devices"[35] such as 'red in the morning, the sailor's warning; red in the night, the sailor's delight.'[36] Although Ong is thinking primarily of language when he considers formulary devices, his emphasis is on standardization with regard to verbal expressions: "other kinds of standardization can be formulary devices in a larger but related sense, insofar as they are matrices for set or standardized verbal expressions."[37] Television programs are not standardized verbal expressions, but as Antin has observed, the standardization of the division of time creates matrices that inform our expectation as to whether the communication will be a program or a commercial. In this way, television's metrically divisible time is a kind of formulary device and is a part of the language of television, likely informing our expectations of what is appropriate digital language as well.

We can also look to examples of hypermedia texts and images to consider the nature of the digital medium and the characteristics of digital visual language, realizing that present examples are "versions of what they are becoming, a structure for what does not yet exist,"[38] and search out the similarities and differences between them and their pre-digital print culture counterparts. William Dickey, for example, an award-winning poet (not to be confused with James Dickey) has published poems extensively in traditional print media for years but has become involves in hypermedia as "an extension of his poetic ideas about the reintegration of image and text." Dickey writes:

> The first important element of these experimental poems is, to me, their effort to avoid the traditional separation of text and illustration, a separation that has governed Western printing since the days of mediaeval manuscripts – I hope we may find a way to think of text and illustration as one continuous field, as it is in the *Book of Kells*.
>
> The second vital element is the avoidance of hierarchical linear organization. Ideally, any hyperpoem can be read in as many different ways as the number of its cards permits.[39]

In the hyperpoem *Heresy*,[40] Dickey uses clip art from MacCalligraphy files and from Dragons by Enzan Hoshigumi Co. and buttons (a visual icon used in HyperCard that allows the viewer/reader to move to another card) from Icon Factory libraries from Hyperpress Publishing Corp. In doing so, he may have avoided the traditional visual language of Western printing that separates text and illustration, but he alludes to the visual language, the compositional wholes, of images that will appear repeatedly in many different cultural contexts, indeed are designed and marketed to move repeatedly from one context to another. In this case, the image is more appropriately *of* digital visual language than the "pencil" line in *Uncle Buddy's Phantom Fun House*, but due to the cultural context – the marketing intention that gave rise to the original image – the image creates a field less like the *Book of Kells* and more like the *Mona Lisa* used to sell pasta sauce.

Clearly there is no clean line between print culture and digital culture, whether one is looking at the visual language or cultural context of the image: the image is both transformed by the medium and transforming the medium. Although some "drawing" in the computing environment is probably necessary and appropriate as part of a digital syntax, drawing as process in this context is changed significantly by the constraints of the medium's hardware and software. This is true despite the uneasy marriage between technology and the old thought idiom – despite the hardware designers with silicone kibbles and silicone graphite trying to bridge the gap between an electronic book and something a dog can eat, the gap between marks that appear on the monitor screen and the hand that puts them there. And perhaps part of the cultural context of the digital medium, such as its tendency towards multiplicity, will embrace the desire to use the same image in many differing contexts, and in so doing so thoroughly change the relationship between the compositional whole of an image and its placement in various contexts that it cannot yet be imagined in this, the late age of print.

In trying to draw such lines, we must be aware that, as Joyce has written, "we have been talking so long about a new age, a technological age, an information age, etc., that we are apt to forget that it is we who fashion it, we who discover and recover it, we who shape it, we who literally give it form."[41] Yet even as we fashion, discover, recover and shape what we are becoming, this print culture attempt at investigating what digital culture is beginning to look like, by nature of print as medium and the cultural context of print, is more clearly about the end of the one than the beginning of the other.

Dana College
Blair, Nebraska

NOTES

[1] Rita Gilbert and William McCarter, *Living with Art* (New York: Alfred A. Knopf, Inc., 1988), pp. 439–42.

[2] Leonardo da Vinci, *Mona Lisa*, Louvre, Paris, Illus. 2.26 in *The Art of Seeing*, by Paul Zelanski and Mary Pat Fisher (Englewood Cliffs, NJ: Prentice Hall, Inc., 1991).

[3] Marcel Duchamp, *L.H.O.O.Q.*, Private Collection, Illus. 531 in *Living with Art* (New York: Alfred A. Knopf, Inc., 1988).

[4] Prince Foods Advertisement, Advertising Annual 1986, *Communication Arts*, Dec. 1988: 17.

[5] Arlene Raven, "Arts: The Archaic Smile," *Ms.* 3.1 (1992): 68–72.

[6] Gilbert and McCarter, p. 420.

[7] Edouard Manet, *Déjeuner sur L'Herbe*. Louve, Paris, Illus, 9.14 in *Worlds of Art* by Robert Bersson (Mountain View, CA: Mayfield Publishing Company, 1991).

[8] H. W. Janson, *History of Art* (New York: Harry N. Abrams, Inc., 1986), p. 14.

[9] John De Andrea, *Manet: Déjeuner sur L'Herbe*, J. B. Speed Art Museum, Louisville.

[10] Jean Fraser, "Celestial Bodies," *Stolen Glances: Lesbians Take Photographs*, ed. Tessa Boffin and Jean Fraser (London: Pandora Press, 1991), pp. 76–85.

[11] Tessa Boffin and Jean Fraser, ed., introduction, *Stolen Glances: Lesbians Take Photographs* (London: Pandora Press, 1991), p. 9.

[12] Boffin and Fraser, p. 9.
[13] Adrienne Rich, *Blood, Bread and Poetry: Selected Prose 1979–1985* (New York: W. W. Norton and Co., 1986), pp. 23–75.
[14] Marshall McLuhan and Quentin Fiore, *The Medium is the Message* (New York: Simon & Schuster, 1989).
[15] Käthe Kollwitz, *Fettered Man*, Private Collection, New York, Illus, 215 in *Living with Art*, by Rita Gilbert and William McCarter.
[16] Käthe Kollwitz, *Liebknecht on His Deathbed*, Märkisches Museum, Berlin, Plate 91 in *Käthe Kollwitz*, by Otto Nagel (Greenwhich: New York Graphics Society Ltd., 1963).
[17] Walter Ong, *Orality and Literacy: The Technologizing of the World* (London and New York: Methuen & Co., 1982), p. 31.
[18] Michael Heim, *Electric Language: A Philosophical Study of Word Processing* (New Haven: Yale University Press, 1987) p. 54.
[19] Ong, p. 33.
[20] Ong, pp. 34–6.
[21] Heim, p. 55.
[22] Michael Joyce, "Notes Towards and Unwritten Non-linear Electronic Text, 'The Ends of Print Culture' (a work in progress)," *Postmodern Culture, an Electronic Journal of Interdisciplinary Criticism* 2.1 (1991): 16.
[23] Stephen Wilson, *Multimedia Design with HyperCard* (Englewood Cliffs, NJ: Prentice Hall, 1991), p. 3.
[24] Joyce, p. 6.
[25] John McDaid, *Uncle Buddy's Phantom Funhouse*, Interactive computer novel (Cambridge, MA: East Gate Systems, Dec. 1992).
[26] Joyce, pp. 7–9.
[27] Nathan Goldstein, *The Art of Responsive Drawing* (Englewood Cliffs, NJ: Prentice Hall, 1992), p. 203.
[28] L. Bernard Happé, *Basic Motion Picture Technology* (New York: Hasting House, 1975), p. 32.
[29] Happé, pp. 31–3.
[30] Arthur A. Englander and Paul Petzold, *Filming for Television* (New York: Hastings House, 1976), p. 121.
[31] David Antin, "Video" The Distinctive Features of the Medium," *Video Culture: A Critical Investigation*, ed. John G. Hanhardt (Layton, UT: Gibbs M. Smith Inc., 1986), p. 156.
[32] Antin, pp. 154–5.
[33] Walter Ong, *Rhetoric, Romance and Technology: Studies in the Interaction of Expression and Culture* (Ithaca and London: Cornell University Press, 1971), p. 285.
[34] Ong, *Rhetoric*, p. 286.
[35] Ong, *Rhetoric*, p. 285.
[36] Ong, *Orality*, p. 35.
[37] Ong, *Rhetoric*, p. 288.
[38] Joyce, p. 5.
[39] Wilson, p. 9.
[40] William Dickey, *Heresy*, Figure 1.5 in *Multimedia Design with HyperCard* by Stephen Wilson.

PART IV

COLETTE MICHAEL

BILLY BUDD: AN ALLEGORY ON THE RIGHTS OF MAN

Shortly before his death in 1981, Herman Melville finished *Billy Budd*, a novel published posthumously in 1924. The action takes place, first aboard the ship "The Rights of Man," then aboard the "H.M.S. Indomitable," at sea, August 1798, the year following the Naval mutinies at Spithead and the Nore, and shows life on board a British warship at the end of the eighteenth century. The most common interpretation of *Billy Budd* sees in this story of two seamen a confrontation of good and evil during one of the many wars that pitted the English against the French.

While there is indeed validity to this reading, a new dimension can be added to this novel by viewing a British film brought to the screen in 1962 by Peter Ustinov who also directed and starred – as Captain Edward Fairfax Vere – in a rather faithful rendition of the book.[1] What the script writers have done is given us an eidetic reduction of the ideas contained in the book where there is little dialogue but many descriptions.[2] A critique of the film together with one of the book bring out a new dimension to the work: sound. By extension, speech becomes the most important and relevant item of the work, although cinema is also a form of writing: the words used are images, and because images are universal, the language is universal; far from reducing the scope of the ideas expressed, it enlarges it because the same ideas can take shape in a multitude of ways, with or without speech, with or without music, with or without colors. These images as a communication media have sometimes proven to be much more effective than mere words. In the book entitled *Billy Budd*, Melville *tells us about good and evil*; in the movie entitled *Billy Budd* Ustinov *shows us about good and evil*. In the process we will see that a new dimension is added to the work.

One of the best examples of the importance of viewing as opposed to reading, in this particular instance, is the way Billy Budd, the foretopman, is introduced to the audience. He personifies innocence, joy of life. The happy sailor is standing high atop the mast but, although in this precarious position, he is carefree and singing a fetching song. The phenomenon is striking: seeing him is akin to admiring a work of

M. Kronegger and A-T. Tymieniecka (eds.), Analecta Husserliana XLII, pp. 251–258.

art. The sight of this handsome young man enjoying his work is an esthetic experience; it implies goodness even before any conversation has taken place with his shipmates of the merchant ship "The Rights of Man." The generating power of these images can be an experience that no book, no article, no words could have produced. But, since the images cannot be brought onto this paper, the description given in the book must be reproduced:

Something about him provoked an AMBIGUOUS smile in one or two harder faces among the blue-jackets. No less unaware was he of the peculiar favourable effect his person and demeanour had upon the more intelligent gentlemen of the quarter-deck. Nor could this well have been otherwise. Cast in a mould peculiar to the finest physical examples of those Englishmen in whom the Saxon strain would seem not at all to partake of any Norman or other admixture, he showed in face that humane look of reposeful good-nature which the Greek sculptor in some instances gave to his heroic strong man, Hercules. But this again was subtly modified by another and pervasive quality . . . above all, something in the mobile expression suggestive of a mother eminently favoured by Love and the Graces; all this strangely indicated a lineage in direct contradiction to his lot. (15)

At the very beginning of this description, we find the word "ambiguous." This ambiguity comes up again when we learn that Billy Budd was a Hercules endowed with feminine qualities, those of the Graces. We are then given a spiritual portrait of the young man:

For the rest, with little or no sharpness of faculty or any trace of the wisdom of the serpent, no yet quite a dove, he possessed a certain degree of intelligence along with the unconventional rectitude of a sound human creature – one to whom not yet has been proffered the questionable apply of knowledge. He was illiterate; he could not read, but he could sing, and like the illiterate nightingale was sometimes the composer of his own song. (16)

Casting Billy Budd as a youth with blonde curly hair, Shirley Temple style, yet manly enough to go topsail – a job which will be shown to be at times deadly – adequately represents the ideas implied in Billy Budd's description.

Soon appears on the horizon a man-of-war, the "H.M.S. Indomitable." The presence of able seamen are requested for the King's navy. Billy Budd must leave his own homeward-bound English merchant ship to be transferred aboard a seventy-four outward bound warship, the "H.M.S. Indomitable". It is symbolic that this merchant seaman, as the transfer takes place, waves goodbye to the "Rights of Man." Was the transfer done as a punishment? Had he committed a crime? "God only knows," replies Billy Budd. No rebellion in his attitude. "As to his enforced enlistment, that he seemed to take pretty much as he was wont to take any vicissitudes of weather. Like the animals, though no philosopher he was, without knowing it, practically a fatalist" (12). Asked if they do not indulge in flogging on the ship the "Rights of Man," Billy Budd, taken aback, can only stutter, "No sir." This suggestion makes him uncomfortable, a fact that Melville paints with poetic evocation: Billy Budd had no "visible blemish"; no, but he had "an occasional liability to a vocal defect. Though in the hour of elemental uproar or peril, he was everything that a sailor should be, yet under sudden provocation of strong heartfeeling his voice, otherwise singularly musical, as if expressive of the harmony within, was apt to develop an organic hesitancy – in fact, more or less of a stutter or even worse" (17).

The stuttering is the sailor's Achilles' heel; he becomes vulnerable when he is excited because he can no longer express himself. And as in a Greek tragedy, this tragic flaw will be his downfall. He stammers because he does not find the words for what he feels. Speech for Billy Budd is a shield and an armor; he is protected as long as he can speak; thus language is a defensive weapon. But as a weapon language – or the ability to speak – can also be destructive as he will soon find out. Routinely Billy Budd is a gentle soul, inclined to love naively the world around him; he meets unfairness with disarming candor and, without any planning or plotting, wins the affection and admiration of other sailors, even the approbation of the ship's officers. His sunny disposition makes for good rapport with his shipmates – all except one, Claggart, the master-at-arms. The latter will do all in his power to tease – verbally – and provoke him, including committing very cruel acts toward other shipmates such as sending a sick man aloft with the excuse that the ship is undermanned. Again in the book *we are told* of Claggart's cruelty; we are told that in him was "the mania of an evil nature, not engendered by vicious training or corrupting books or licentious living, but born with him and innate, in short, a depravity according to nature" (46). In the film *we are shown* Claggart's cruelty; he orders a sick man to go up the mainmast, although the feverish sailor can barely stand up. Refusing to obey the order would have been mutiny, the sailor therefore goes up and falls to his death. The incident shows how a man is killed because of the unfairness and evil of the master-at-arms.

Since they are sailing under the articles of war, there seem to be only two duties: obey orders and fight (the enemy). Claggart appears to be afraid to let his guard down and become fond of someone; presumably because too much of the navy is made up of thieves and murderers but more to the point because "some men cannot stand too much perfection." It is a kind of disease.

Claggart has a most pleasant conversation (in the film) with Billy Budd, and, for a short while, one may feel that the fact that Billy Budd in attempting to speak to this lonely man will break the barrier that stands between himself and Claggart, that speech will be the weapon that will disarm Claggart. Topside Billy Budd simply asks permission to watch the water but he does not appraise at its right value the sarcastic reply of the master-at-arms, who "suppose(s) the Handsome Sailor may do many things forbidden to his messmates" (30). And symbolic of the two men's view of the world is their respective evaluation of the sea: to Billy Budd, the sea is "calm and peaceful". To the master-at-arms, it is deceitful, "calm above, and underneath, a world of gliding monsters preying on their fellows. Murderers, all of them. Only the sharpest teeth survive" (30). In the sea both find a reflection of their own nature, one gentle and at peace with himself; the other tortured and cynical. But Clagggart cannot repress a certain curiosity about this Handsome Sailor and asks him:

But tell me this: how have you stomach to stand here and talk to me? Are you so innocent and ignorant of what I am? You know my reputation. . . . Have you not intelligence enough to be afraid of me? To hate me as all the others do. (31)

Billy Budd does not understand why he should be afraid and does not quite understand why some of the men are. To which Claggart replies "You are a fool, fellow. In time, you'll learn to fear me like the rest . . . " (31). In his answer Billy Budd is projecting his own goodness in others and refuses to be insulted for being called a fool. Instead he assumes that being a fool means having no formal learning and he states:

Oh, I've got no education, I know that. There must be a lot of things a man misses when he's ignorant. But learning's hard. Must be sort of lonely, too. (31)

Intuitively Billy has realized that knowledge may set a person apart, perhaps that knowledge might even, at times, be a burden – indeed, this will prove to be true for Captain Vere – and Billy has a special insight which is lacking in Claggart, who replies:

What are you prating of, half-man, half-child? Your messmates crowd around, admire your yellow hair and your blue eyes, do tricks and favors for you out of love, and you talk about loneliness! (32)

Mixture of envy and desire. Jealousy perhaps. There is in the young sailor something that Claggart will never have. That longing was noticed by the sailor as Claggart was looking off to leeward: "Kind of sad, you were looking" (32)

Claggart can only deny this probing and replies: "Not sadness, boy. Another feeling, more like . . . pleasure. That's it. I can feel it now, looking at you. A certain . . . pleasure." (32)

Billy Budd should have been repulsed by this uttering, but he failed to understand that he was going to pay for having innocently seduced the master-at-arms. Instead, looking very flattered, he can only thank the master-at arms who, then, wants to know if the sailor "likes" him!

Although Billy Budd is naive and innocent, he is not pictured as lacking intelligence. Quite the contrary. His repartees shows a sensibility and certainly a quiet empathy for his fellow men and he possesses, if not a formal knowledge of the world, at least of the heart. This he even recognizes; he knows he has no education and states as much. But his youthful beauty and unsophisticated good-fellowship, as pointed out by E. L. Grant Watson,[3] exercise a consequent repulsion on Claggart and "make it only easier for these qualities to turn envy into hatred. His [the sailor] very virtue makes him the target for the shaft of evil, and his quality of acceptance provokes to action its complementary opposite, the sense of frustration that can not bear the consciousness of itself, and so has to find escape in mania." Claggart could not cancel the evil in him: he could not let himself go to loving and caring in a normal way for this "moral phenomenon" who was the Handsome Sailor. There was only one recourse left to a nature like his, "surcharged with energy as such natures almost invariably are, but to recoil upon itself, and, like the scorpion for which the Creator alone is responsible, act out to the end its allotted part." Claggart will increase his verbal abuses against Billy Budd. "Thus there develops the conflict between unconscious virtue and the bitter perversion of love which finds its only solace in destruction" (Grant Watson, 15).

The conversation barely ended, Claggart gives the order to put Billy Budd on report, on any ground whatsoever, real or invented.

One of the sailors who does not have Billy Budd's innate goodness will attempt to kill Claggart; he is stopped by Billy Budd who talks him out of it. The young man uses his power of speech, of communication, to avoid and stop evil. But Claggart reports to the Captain a fake story – a verbal assault on the young sailor's integrity. Again this demonstrates how words, if words can save, can also destroy a man. Claggart destroys Billy Budd's reputation with such an incredible lie that the young man loses his composure and his power to defend himself, the power of speech; he is clearly in a state of shock; he cannot express himself; yet, he feels he must react in the only way he knows under such a provoking situation; he reverts to body language, to an uncivilized reaction, the reaction of force and he strikes the master-at-arms, inadvertently killing him.

The focus then switches to the captain of the ship and other officers. Although Billy Budd was provoked and the blow was accidental, the sailor must be court-martialed on board; in effect the living must stand trial for the dead man's crimes. The ship's officers assembled for the court-martial all like Billy Budd. They all realize that the sailor was provoked beyond endurance; they can even see that there existed an element of self-defense in his reaction but, again, we witness this court-martial as a decision made on the best speeches. The original vote gives the first verdict: not guilty.

Here we see that there are two laws as pointed out by several critics: "The civil code – represented by the Articles – by which Vere and the members of the court are bound as officers, and the law of nature – romantically identified with the heart – in the light of which Billy is innocent.[4] That was the first impulse, the first reaction, one dictated by the heart. But then the Captain will speak: he is in a tough position as an officer. He has taken an oath to obey a code that made him an officer; in time of war can he condone such an act as the one committed by Billy Budd? "Though he believes in Billy's innocence, naval law and prudence alike demand punishment for the impetuous seaman while pity and reason counsel mercy. Internal debate inclines the captain toward conviction."[5] The law is conceived to serve justice, but acquitting the sailor is beyond the scope of military justice.

> As officers we are concerned to keep this ship effective as a weapon. And the law says what we must do in such a case as this. Come now, you know the facts, and the Mutiny Act's provisions. At sea, in time of war, an impressed man strikes his superior officer, and the blow is fatal. The mere blow alone would hang him, at least according to the Act. Well then, the men on board know that as well as you and I. And we acquit him. They have sense, they know the proper penalty to follow, and yet it does not follow. (73)

Somehow, in a blatant contradiction, only the so-called civilized man's acts of violence can conserve order and peace. Somehow the officers must punish the sailor, and the Captain states:

> The world demands that we keep an order we cannot understand. That's true. The world demands it: demands that at the back of every peace-maker there be the gun, the gallows and the gaol. I talk of justice, and would turn the law gentle for those who serve here; but a Claggart stands in my shadow, for I need him. So the world goes, wanting not justice, but order . . . to be let alone to hug its own iniquities. Let a man work to windward of that law and he'll be hove down. No hope for him, none.

Captain Vere and the other ship officers, if they want to abide by the Mutiny Act's provisions, in order to keep the ship as an effective weapon, must condemn a man who struck a superior officer. Billy Budd is after all found guilty of murder and must be hung until dead. So that even under these conditions – a war – it is not a firearm that will destroy Billy Budd but again a speech:

> But we are public men, controlling life and death within this world at sea. Tell me whether or not in our positions we dare let our consciences take precedence of the code that makes us officers and this case to trial. (77)

The Captain may feel revulsion and shame, but he must enforce the law and condemn the "luckless boy." He cannot let his "conscience take precedence" over the code that makes him an officer. In an impassioned speech to his officers, Captain Vere rationalizes the choice he forces upon them:

> Do not think me pitiless in thus demanding sentence on a luckless boy. I feel as you do for him. But even more, I think there is grace of soul within him that shall forgive the law we bind him with, and pity us, stretched on the cross of choice. (78)

Melville's story is certainly about a confrontation of good and evil but it is a special kind of evil, produced solely with words, with cruel and unreasonable orders, with words that taunt and provoke to the point where one loses control of the only defense acceptable, when it becomes impossible to utter words, when one is so shocked as to be speechless. It is also gratuitous evil, not done with intent to gain, but simply with intent to annihilate. What is extraordinary about this novel is that it shows that words can be used for a dual purpose, to soothe, to calm, to defuse, but also to hurt mortally as surely as with a loaded gun even if done with the blessings of the law.

Billy Budd has the last word and with it, the only victory, a moral victory for, as he is about to die, sure of his innocence, he shouts: "God

Bless Captain Vere." In time of war, and because of the state of war, a seaman must die, not in battle, but killed by words. There is a corollary to this finding; again it is demonstrated that there can be no absolute in morality.

Northern Illinois University

NOTES

[1] The movie script is an adaptation of the dramatization by Louis O. Coxe and Robert Chapman (New York: Hill and Wang, 1962). All the quotes in this article are from this Broadway adaptation originally written in 1947 and revised in 1951.

[2] There has been many editions of *Billy Budd*, which was begun by Herman Melville on Friday, November 16, 1888, revised from March 2, 1889 until it was finished in April 19, 1891. Used here is the standard edition of *The Works of Herman Melville*, edited by Raymond W. Weaver (New York: Russell & Russell, 1963).

[3] E. L. Grant Watson. "Melville's Testament of Acceptance, " *Studies in Billy Budd* (Columbus, Ohio: Merrill Publishing, 1970), p. 14

[4] Walter Sutton. "Melville and the Great God Budd," *Studies in Billy Budd* (Columbus, Ohio: Merrill Publishing, 1970), p. 86.

[5] William York Tindall. "The Ceremony of Innocence," *Great Moral Dilemmas in Literature, Past and Present*, edited by R. M. MacIver (New York: Harper and Row, 1956).

KENNETH W. BREWER

"POLITICAL ALLEGORY: BAKHTIN, JAMESON AND BIRAGO DIOP'S *LES CONTES D'AMADOU KOUMBA*"

The marginalization of oral literature by Fredric Jameson and other Marxist literary critics is a surprising phenomena. Oral literature is given minimal attention, and it almost always plays the role of the underdog, the marginalized voice suppressed by the hegemonic powers of writing. The privileged position of the written word and of the traditional definitions of genre have historically excluded oral literature from the canon. This might be an indication of a general level of traditionalism and conservancy in Jameson's and other Marxists' critical approach. A look at the question of the validity of the appropriation of certain of Mikhail Bakhtin's critical concepts by Jameson brings attention to some of the polemics that are centered on the dialectic, language, history, and genre and are relevant to the discussion of oral literature and to Marxist criticism in general. Birago Diop's *Les Contes d'Amadou Koumba* (Paris: Présence Afraicaine, 1961) is a compilation of oral tales that provides a decent sampling of the linguistic, allegorical, and generic diversity that one can find in oral literature; with it we can see more clearly how Jameson fails to deal with this diversity.

Fredric Jameson's reading of Bakhtin has been criticized as a Marxist appropriation *par excellence*. Much of this criticism is based on Jameson's dialectical readings of Bakhtin. In the first chapter of his book *The Political Unconscious: Narrative as a Socially Symbolic Act* (Ithaca: Cornell University Press, 1981), Jameson pulls Bakhtin's concept of the dialogic into his marxist paradigm. It is done so overtly:

> . . . it will be necessary to add the qualification that the normal form of the dialogical is essentially an antagonistic one . . . Within this new horizon, then, the basic formal requirement of dialectical analysis is maintained, and its elements are restructured in terms of contradiction (this is essentially . . . what distignuishes the relationality of Marxist class analysis from static analysis of the sociological type) (*PU*, 84–85).

Bakhtin specifically rejected the Hegelian dialectic. This would appear to contradict Jameson's placing the concept into his "new horizon". Jameson however, is consistent insofar as his inclusion and interpretation of Bakhtin parallels his lengthy analysis of Althusserian Marxism.

M. Kronegger and A-T. Tymieniecka (eds.), Analecta Husserliana XLII, pp. 259–264.

The Althusserian dictum, "History is a process without a *telos* or a subject" [*Réponse à John Lewis* (Paris: Maspéro, 1973), 91–98] is seen, by Jameson, as a modification of the dialectical tradition rather than a rejection. Jameson thus implies that Bakhtin's analysis of literature in light of history and social conflict automatically has a Marxist bent even if it is not tied to the utopic ends of traditional Hegelian Marxism.

More than one critic has claimed that this Marxist reading of Bakhtin by Jameson is based not on a misunderstanding of Bakhtin, but rather on Bakhtin's application of dialogism as opposed to its theoretical tenor. This seperation of theory and praxis, however, is not all together clear since Bakhtin uses a diverse set of literary examples, to construct his theory.

For Bakhtin, the role that the folktale plays in the production of the novel is subversive; however, the results are very positive. The "absolute past", "epic distance", "language of the dead", and all that which is "inaccessible to personal experience and does not permit an individual, personal point of view or evaluation" (*Dialogic Imagination* 16), is somehow violated and injected with contemporaneity by folklore and oral traditions. The important role that oral literature plays is only in the pre-history of the novel and is eventually replaced by "parodic-travestying" literary doubles – intertextual antagonism. Bakhtin's dialogue with the Russian formalist critics produced an emphasis on everyday speech which these literary forms – oral and written – helped incorporate into the novel. Bakhtin's analysis is insightful but it lacks a deep look at the reverse relationship between written genres and the folktale and the parallel effects of parody on oral traditions.

One of the stories in Birago Diop's collection, "Vérité et Mensonge" ("Truth and Falsehood"), can provide here an example of how an oral tale might produce an equivalent dialogic dynamic within oral traditions. "Vérité et Mensonge" is a parody on the moralizing fables that one finds in almost every culture on the virture of telling the truth. Vice and virtue are personified in the story (Fène-le-mensinge and Deug-la-vérité) in a way that is not dissimilar to the medieval morality plays. As the two characters decide to travel together, Fène-le-mensonge decides to keep silent for fear of getting in trouble. After several mishaps that are caused by Deug-la-vérité's inability to lie, Fène-le-mensonge finally speaks and is able to trick a king out of his kingdom. The comic outcome of this story can be seen as being subversive to the religious leader, moralizing storytellers and the oral form it is parodying. The form of the oral tale being parodied might be as static as the epic or poem which play an essential role in the production of the novel in Bakhtin's system. The 'centrifugal' force that disrupts the dominant, authoritarian, and monologic oral tradition, works the same way as it does with writing and form.

While criticizing the formal approach to 'living' heteroglossic work, as opposed to those works which no longer are dynamic in form and meaning, Bakhtin failed to show how the parallel dynamics work in oral literature. Perhaps the failure of Bakhtin here is simply the fact that one of his larger projects was a history of the novel, and that while very non-traditional, he had to maintain a certain level of consensus on what a novel is – a written work – to advance his arguments against the current system. Another possibility is that Bakhtin, like many of his contemporaries as well as many critics today, shared the belief that political power was directly tied to literacy. This power relationship, while valid in some cultural and historical contexts, is not universally true in, for example, non-literate or pre-literate societies.

In *The Political Unconscious*, Jameson describes genre as having both a semantic and syntactic (or structural) element. These two elements correspond to Bakhtin-Voloshinov's "two trends of thought in the philosophy of language" (*Marxism and the Philosophy of Language*, 45–63). Jameson incorporates both of these two aspects of genre into dialectical historical "superstructure":

> Dialectical thinking can be characterized as historical reflexivity, that is as the study of the object . . . which is also the study of the concepts and categories (themselves historical) that we necessarily bring to the object (109).

Here we find a divergence in the two critics' approaches. Jameson does not make the same distinction between the 'living' and the 'dead' language in literature and so his criticism of structuralism is based on the lack of historical reflexivtiy rather than this dichotomy of language. Historical reflexivity, in essence, makes all works 'living' and relevant. Bakhtin rejected any type of synthesis on the grounds that these two "trends" are processes of abstraction, a process that creates misunderstanding and that would eliminate the possibility of their proper synthesis. Jameson, on the other hand, wants to formulate a structuralism with meaning and believes that history, rather than being an uncontrollable wild card as it is for Bakhtin, is the final link missing in his ultimate structural system. Jameson's analysis of the deficiencies of the formal/structural systems of Vladimir Propp [*Morphology of the Folk Tale*, trans. L. Scott (Austin: University of Texas Press, 1968)] and of Claude Lévi-Strauss (*Mythologiques*) in dealing with the human character results in a historical tie between form and the notion of the "pre-individualistic narrative".

In his article entitled "Third World Literature in the Era of Multinational Capitalism" [*Social Text*, no. 15 (Fall 1986)], Jameson develops this concept of the "pre-individualistic narrative" and relates it directly to the questions of modernism and high realism. Political allegory, or what Jameson terms "national allegory", is the inescapable "collective" spirit that is the essential characteristic of "third world" literature. Realism and modernism, the products of the enigmatic "first world", result out of the "privatization" of language that occurs in the post-feudalist capitalist modes of production. This "privatization" of language produces a discrediting of allegory and the separation of the poetic from the political – "Freud vs. Marx." The oral folktales or "pre-individualistic narrative":

> . . . emerge from a social world in which the psychological subject has no yet been constituted as such, and therefore . . . later categories of the subject are not relevant. Hence the bewildering fluidity of these narrative strings, in which human characters are ceaselessly transformed into animals or objects and back again . . . nothing like narrative "point of view," let alone "identification" or "empathy" with this that protagonist emerges (*PU*, 124).

The emphasis on the interchangeability of these narrative types deemphasizes their distinct cultural roles. Granted, the length of most folk tales does not allow a great deal of character development, however this may also be said of the short story. And if "empathy" is the goal of the "individualistic narrative," then the separation between the political and the poetic that Jameson claims is characteristic of the "individualistic narrative" is undermined. The importance of allegory in oral traditions is clear, but the existence of non-allegorical elements in oral tales should not be overlooked. In Diop's story "Un Jugement", for example, we can find meaning in both surface and deeper allegorical readings. The main character, Demba, repudiates his wife (Koumba), and almost immediately afterwards regrets it. He then denies that he repudiated her as a ploy to force her return. The wife, enjoying her life as a single woman, refuses to go back to him. They are forced to seek mediation, and after visiting several villages in search of a resolution, Demba is finally tricked into admitting that he did repudiate his wife by a village *iman*. We could read this whole story as a simple "know a good thing when you have it." However, as the main characters have a certain level of verisimilitude (not doing anything out of society norms, no super-human feats, etc.), and there is some character development (Demba reflects on his actions several times during the narrative), the issue of divorce and of certain types of behavior that appear on the surface of the text might be of the greatest interest to the listener in the environment in which the story was told.

Diop's selection of stories in *Les Contes* includes bestial, parodic fables, and somewhat realistic narratives with mostly human characters, such as "Le Jugement." This suggests that Jameson's formalization of the "pre-individualistic narrative" and its tie to oral literature is not entirely valid.

The next jump for the "third-world," made possible by the supposed bypass of feudalism in Africa and China, is to the "post-individualistic narrative." This "post-individualistic narrative" like its predecessor, is based on the primacy of allegory. "Post-individualistic narrative" differs from "pre-individualistic narrative" only in complexity. Allegory here, as opposed to a one to one correspondence, "generate(s) a range of distinct meanings or messages simultaneously" ("Third World", 148). Jameson's analysis of the "post-individualistic narrative" is exclusively based on written works. It is certain that none of the stories in *Les Contes d'Amadou Koumba* are anywhere near the ultimate frontier of the allegorical or structural complexity that is possible in an oral tale. Writing does allow a level of correction and re-working that is difficult to equal with memory alone, but the limits of the human intellect are rarely explored and there is no reason to believe that "deep", polyvalant or complex meaning is intrinsic or exclusive to writing. Jameson ignores the problems of interpretation and reading which further complicates matters.

Jameson's use of the "first world"/"third world" dichotomy makes the attemp to link allegory to history and genre even less credible Furthermore, by tying allegory to genre and history, Jameson is apparently just refining the struturalist approach and violating the very essence of the Bakhtinian project.

Bakhtin himself was more aware than most people of the problems of abstraction and of classification. Who qualifies as a Marxist critic? and, What do we call literature? have by no means been answered here. We have seen, however, how both Jameson and Bakhtin marginalize oral literature, each in their own way and to a different degree. Whether Bakhtin was a Marxist literary critic or not appears to be more related to the question of the need for a historically formulated structural paradigm for Marxist criticism like the one Jameson promotes. The marginalization of oral literature, however, is based more on the historical precedent in literary criticism than on its structural consistency. The re-evaluation of oral literature in Marxist criticism will take a revolutionary rethinking of what literature is, as well as what approach to take in its analysis.

Pennsylvania State University

BERNADETTE PROCHASKA

THE MYTH OF THE FALL AND WALKER PERCY'S WILL BARRETT

Adam is wandering in a land outside a lost garden. He is bewildered by the signs of the times. Something is amiss in the world and in his personal life. Science is of no avail in his search for personal health. He is a fallen man in a fallen world. His story is the enduring story of the Myth of the Fall as it is found in the Book of Genesis. Walker Percy's Williston Bibb Barrett, in both *The Last Gentleman* and *The Second Coming* is Adam on a journey, conscious of the fact that secrets exist all around him in a post-modern world, where in Percy's own words: "men discovered that they no longer could understand themselves by the theory professed by the age" (MB 25).

As an allegory, the Myth of the Fall incorporates dual interests. In its literal sense, the story in the Book of Genesis tells of a relationship between God and man which undergoes a serious catastrophe. Adam and Eve lose a splendid life they had experienced with their maker and are driven from the Garden of Eden. Present in the simple story of the Fall is a marvelous and enduring beginning of the reality of salvation history. For as the Myth of the Fall incorporated the terrible and omnious statement "and you will die," so it also incorporates the wonderful promise of hope when God says to the serpent: "I will put enmity between your seed and her seed, and she shall crush your head and you will lie in wait for her heel" (Genesis 3;15–19). The "seed" of Eve includes all humanity including Christ, whose birth, death and resurrection is the hope of redemption for the human race. The Myth of the Fall is then the beginning of salvation history which will endure until the Parousia. It is the story of death and resurrection; it is the story of an infirm world inhabited with infirm individuals who can look forward to the promises which Christ has won by his Redemptive life.

In Percy's novel, *The Last Gentleman*, Will Barrett is Adam as a sovereign wayfarer on a journey from "ground zero" in Central Park to Santa Fe where he is companion to a young man facing death. He participates as interpreter for the Chaplain, Father Boomer, in the sacramental salvation of Jamie Vaught, the dying boy. The death scene follows a long spiritual sojourn for Will. On the literal journey, which

M. Kronegger and A-T. Tymieniecka (eds.), Analecta Husserliana XLII, pp. 265–276.

compromises most of the novel, he is bewildered in his encounters with members of the Vaught family; with Kitty who would be his sweet love except that he cannot understand her unpredictable whims, with Rita who would tempt him to the inaction of self-satisfaction, with Sutter who would exit the struggle because it has no meaning for him. However, with Valentine Vaught, Will participates in seeing that Jamie is baptized. He thus experiences the meaning of the Fall as death, while at the same time he is introduced to the meaning of hope in the Resurrection. Will encounters death in the horror of the hospital room where it has made its presence known in the terrible stench of decay. He encounters the hope of resurrection, when Father Boomer, reading the ritual for the dying, says to Jamie, "Today you shall see God." The first novel about Will Barrett is a story of the journey of a soul, Adam engaged in the terrible punishment for sin, yet engaged in the wondrous certitude of redemption.

In *The Second Coming* Will Barrett is the new Adam literally falling on the golf course. The allegory of the fallen man in a fallen world persists. Will recognizes his own confusions and tendencies toward self-destruction, as well as the malaise and manipulative machinations of those around him. Why is everything out of joint? Sometimes it is in a wrong word, as in Allison's speech; sometimes it is in an environmental "outcropping" as with the stones on a hillside, and sometimes it is in misplacement of loyalties, as in the case of Leslie and Jack Curl. His Lost Paradise of a world includes an out-of-bounds thicket into which he uncontrollably hits his golf balls, a ruined greenhouse into which he falls, a swamp where his father once tried to kill him, and a cave in the bowels of the earth where he goes to meet God. His story is one of salvation for he moves toward integration into "communitas" in his love for Allison, who has inherited the ruined greenhouse where Will experiences a rebirth. Most notable of his actions is his participation in building a village of garden homes. His desire to work with others is a means by which he discovers his own identity. Adam, bungling all the while, remembers that there must be some glorious place where God once walked with Eve and himself, forming the first community. Will Barrett moves toward a similar community by insisting that both Father Wetherbee, an old missionary priest who believes in Apostolic Succession, and Allison share his life. Participation as integration is Will's salvation; it is his movement toward renewing the face of the earth and renewing himself. Vanda orchids will bloom in the greenhouse tended by an old gardener discovered by Will. Something which has been lost in nature will be recovered.

In the first Will Barrett novel, the allegorical opening sentence "One fine day in early summer a young man lay thinking," reads like the beginning of a fable or a fairy tale. The man is a "young" man, a new Adam, who in the first fourteen pages of the tale doesn't have a name. Being nameless, he could be anybody, an "everyman", lying on the ground in Central Park, a garden of sorts, a place somewhere east of Eden, in a post-lapsarian fallen world. In his article, "Percy and Place", John Hardy notes that the place where we first meet Will is a place where no one is at home (8). Percy deliberately delays the start of his hero's adventures by "displacing" him. It is early summer when we first meet Will. The great meadow is filled with life on a fine day in the glorious season of growth. However, something is wrong. The young man, still nameless, is thinking about his youth, how he had lived in a state of liveliest expectation. What a fine thing it would be to become a man and know what to do, like the Apache youth who at the right time goes out into the plains alone, dreams dreams, sees visions, and returns, knowing he is a man (LG 11).

Becoming a man and knowing what to do is an ancient endeavor, the task of Adam's sons. For Will, no such time had come. Although he had hoped for some illuminating allegorial experience which he cannot name, it has not happened. He does not have the awareness of an enduring transcendent presence. Will is, in fact, a man of the "post-modern age". Percy writes that "the modern age began to come to an end when men discovered that they could no longer understand themselves by the theory professed by the age" (MB 25). Because it allies itself to a rational theory which does not transcend the material world, modern science cannot produce satisfactory answers for the human spirit. The time, unencumbered by science, in the life of the Apache who could find his identity associated with his dreams and visions, has still not come for Will Barrett. Will listens with his good ear, and he uses a telescope to help him see more clearly. Nevertheless, technology and the scientific theory which is based on careful observation will not satisfy him. Uneasy about his situation, he is the modern man who does not know how to pose the question.[1] Will senses that there is a great secret somewhere beyond his knowledge, and he uses whatever means he can to discover that secret, the chief means being the telescope and eavesdropping.

Percy's protagonist buys a telescope so that he might see details more clearly. It is his analyst, Dr. Gamow, who asks if his client intends "to become a seer". Will's reply is that of Adam the seeker who finds that technology could be his salvation: "Ha-ha, there might be something to that, a see-er. Yes" (LG 37). Will becomes the typical creature created by the god, science: the objective observer of society in all its detail, a soulless entity who sees surfaces and stops there, the gnostic who discovers the "answer" in the material world. In referring to the impulse of modern science, Cleanth Brooks has remarked that the scientists "have erased the notion of the Fall and have promised to take man back to Eden" (54). In an effort to take his patient back to Eden, Gamow represents Percy's theory of science as it is expressed in the "Delta Factor": "The scientists were saying that by science man was learning more and more about himself as an organism and more and more about the world as an environment and that accordingly, the environment could be changed and man made to feel more at home" (MB 45). However, no such thing has happened for Will, who remains isolated and homeless.

When we discover more about Will Barrett, we find that he is living in New York and working beneath Macy's store as an engineer, watching dials which control the flow of air for the store. A network of unplanned incidents and bungling encounters brings him into contact with the Vaughts, a southern family in the north, to seek a cure for the youngest member, Jamie. Again in a garden, Will sets up his telescope to watch a Peregrine falcon, a pilgrim bird. The bird, Will's alter ego, is a wayfarer watched by the pilgrim bird-watcher:

> And every morning the pilgrim bird patrolled the cornice, making an awkward sashay in his buff pants, cocked his yellow eye at Misty trees below, and fell like a thunderbolt, knocking pigeons out of the air in all directions. The engineer took a dozen photographs in magnification one fifty, trusting that at least one would catch the fierce eclipsed eye of the falcon. (LG 45)

Walker Percy claims that the idea for the marvelous Peregrine Falcon comes from his reading of Gerard Manley Hopkins "The Windhover."[2] In the Hopkins poem, the "dapple-dawn-drawn Falcon" is a symbol of Christ, and in the narrator's telling of the story of the last gentleman, the Falcon is a "sign". The Falcon leads Will to the Vaught family and into his own great journey from Ground Zero to Santa Fe. Turning the "sign" associated with the Falcon into a cosmic and comic situation, Percy engages Will's own theory of immanence, namely: something great and terrible is about to happen or has already happened. Back in his room in the Manhattan YMCA, Will Barrett sees a map of Ground Zero on the front page of the *Times*. It is one of those maps illustrating the effects of the latest weapon, in this case some kind of nerve gas. The innermost circle, called the irreversible axon degeneration, is at the bench in Central Park where he had seen Rita Vaught while he was watching the Falcon. Ground Zero is also a sign. The fallen world in which Adam finds himself is unmistakenly a profoundly perilous place.

Although Will senses danger in the garden, the author, in an elaborate attention to setting, enables his protagonist to see signs and wonders which both give movement to the plot and move the engineer to take the risk, to set off on his journey. The relationship of Williston Bibb Barrett and Jamie Vaught begins in a hospital in New York and continues through a journey which takes them to a hospital in New Mexico where Jamie dies.[3] The journey of Will to Santa Fe, Holy Faith, includes his loving interludes with Kitty Vaught and his frightening encounters with Sutter and Rita. Valentine Vaught, however, is the individual who most markedly engaged Will's participation in the process of salvation.

She tells Will that she believes the "whole business: God, the Jews, Christ, the Church, grace and forgiveness of sins" (301). She claims that while she is motivated by her religious beliefs, she is "meaner than ever", a "good hater" who hopes her "enemies fry in hell" (301). It is the recognition of her fallen human condition that moves Val to look beyond, to rely on grace, and to participate in the life of Christ already begun on earth. Baptism is foremost in her mind when she thinks of Jamie, because she wants him to participate in eternal life which she knows exists. Actually Will speaks for Val, for although he is the one who asks Father Boomer to baptize the boy, the priest tells the youth that baptism was requested by "his sister". Adam, alienated, alone at the beginning of *The Last Gentleman*, shares an important place by his dying friend. By carrying out the commission of seeing that Jamie is baptized, a new Adam is engaged in Christ's work of bringing life to others.

Valentine Vaught has joined a religious order and is known as Sister John Vianny. A central question is raised by Mr. Vaught: "Do you think the good Lord wants us to do anything unnatural?" (84) The concern of the rich father regarding the "unnatural" is an ancient concern. Chandler Vaught cannot understand the identity Val is seeking through what she does with her inheritance. In the past he was the creator of the lives of his children, the supreme provider of their needs. He claims that Val was his "girlie," his child with an infirmity:

When she was little, she used to have growing pains. I would hold her in my lap and rock her in the rocking chair for hours. (84)

Ordinarily children do not suffer from "growing pains". There was something different about Val, according to her father, even as a child. Mr. Vaught is telling Will that he is the provider and protector, revealing the notion that he is a rather diminished great god who could not control the infirmities of his children. But now (and Val is the worst), they have "messed him up".[4]

Walker Percy's second novel about Will Barrett opens with signs and symptoms. Using the word "sign" in the first sentence, the author reveals Will's awareness regarding his own fallen condition: "The first sign that something had gone wrong manifested itself while he was playing golf." Life is good for the middle-age, successful attorney, but poor Will has a falling sickness. Moreover, when he looks at the clouds, he sees shapes like the atomic bomb over Hiroshima. The sun passing over the red maples looks like fire. A little cosmos in itself, the golf course is marvelous testimony to the scientific challenge of the human intellect to control nature. Measured and manicured and carefully nurtured, the park-like area is found in developed countries, where herbicides and pesticides and money-making caretakers produce the wondrous garden-like grounds more beautiful than nature itself can provide. However wonderful it is, an out-of-bounds thicket, like John Chrysostom's allegorical dragon, lies close by to receive the players' misdirected balls.

Besides the golf course, the other important places in *The Second Coming* are the greenhouse, where Allison lives, the cave, where Will goes to learn a secret, the swamp, where his father tried to kill him when he was a boy, and the village of garden homes that Will and Allison plan to build at the end of the novel. Continuing the movement of the first novel, Will embarks on a journey, this time to the bowels of the earth, where he wants to know the answer, yes or no. Although the golf course is the allegorical post-lapsarian Eden, the fallen world, it nevertheless retains a splendor and glory, because of what is hidden within it. Emphasizing its ancient riches, Percy describes it as a mythical haven for a sacred animal: "A thousand feet below the sunny golf course links, the tiger had crawled into the cave two thousand years ago, lain down in the darkness and died."[5]

Throughout the novel, the golf course is an important central locale, for it is adjacent to the greenhouse into which Will falls. He also journeys into the cave for his encounter with God from the Number Fifteen fairway. After his misadventure in the cave, and his confinement at St. Mark's Convalescent Home, Will comes back to the golf course to get Allie so they can be married.

If the journey into the cave is a quest for faith, Will as God-seeker meets his first challenge in his encounter with the tiger. To begin with, he uses scientific knowledge to conclude that the calcareous formation is only the likeness of the tiger. Next, he confidently situates himself in the tiger's den to await the word from God.[6] Unfortunately, the profound question of faith becomes secondary when Will gets a toothache which makes him sick. In an interjection of his own voice, which quite deliberately restates Sartre's description of modern man, Percy notes:

> There is one sure cure for cosmic explorations, grandiose ideas about God, man, death, suicide, and such, and that is nausea. I defy a man afflicted with nausea to give a single thought to these vast subjects. A nauseated man is a sober man. A nauseated man is a disinterested man. (SC 213)

In his sickness, Will is the disinterested man, he is an allegorical everyman-figure representing the malaise of modern man. Grandiose ideas about God, man and death are secondary to his preoccupation with himself. Until he recognizes his infirmity, nothing really happens in the cave. Knowing his infirmity is his turning point. It is the ground point of his presence in the cave from which he will rise spiritually. With the blurred vision brought about by his sickness, Will sees the tiger standing in the opening of the cavern. Will notes that the tiger is too tired even to unlock his legs and lie down. It is clear to him that the tiger had come to die (SC 221). In noting the infirmities of the tiger, its sickness, blurred vision, inability to stand, the reader somehow senses that it bears a close resemblance to Will. Ultimately, the great sickness of Will, like the nausea of modern man, is the ancient presence of original sin. What Will sees in the sick and aged tiger which is unable to stand, is his own condition.

The allegorical Myth of the Fall is a story of alienation, of individuals, who (like the tiger), choose to withdraw from the community which existed in the Garden of Eden where God walked in the evening in the company of Adam and Eve. In their banishment from the Garden because of some unwillingness to obey God, Adam and Eve, and all their descendents, share the condition of alienation, of original sin. Integration into "communitas" is the way of salvation. It is the profound knowledge Valentine Vaught had in her persistent determination in *The Last Gentleman*, to have her dying brother, Jamie, baptized. She knew that baptism is the magnificent initiation into the community of faith which is the family of God.

Most notable of the uses of the golf course is the fact that it lies over the marvelous cave which includes an air passage becoming a vent for the greenhouse built on the side of the mountain. A comic situation occurs in this place created by the author to locate Will's own rebirth, his falling into life after his adventure with death in Lost Cove Cave. With Will's crashing through an old air vent and landing unconscious on her floor, Allison's "only thought had been to make some sense of it". He was a problem to be solved. Her first sense that she indeed had some knowledge of her new world was that she recognized him. It seeemed he was neither dead nor badly hurt. The author notes that "He smelled of a freshly dug ditch. A grave." (SC 234)

Visually, the scene is that of a fallen man covered with blood and clay while he smells of the grave. The place where he lies is the floor of a ruined and neglected greenhouse. Edenic in its very essence, a well-kept greenhouse is a wondrous place where nature not only reaches its own supremecy, but the glass structure contains the colorful blossoming flowers, the result of human labor, the epitome of visual beauty and pleasant fragrance, the glorication of nature. The Garden of Eden was never so beautiful as a carefully tended greenhouse, and the author might have smiled when he created the scene of Will Barrett crashing into the now dilapitated ruin where Allison is on her hands and knees cleaning the floor.[7] The fact that the greenhouse lies in ruins only emphasizes the infirmities of the two individuals who dwell within its breakable glass walls.

When he creates the contrived situation of Will Barrett's fall into the greenhouse, Percy is giving evidence of the association of a physical and mythical resurrection with the spiritual rebirth taking place in the souls of both the major characters. While Allison's rebirth has to do with her recovery of herself through language, Will's movement toward some satisfactory hold on life is more ambiguous.[8] He is post-lapsarian Adam fallen through the vine-covered air vent into the ruins of the greenhouse attached to a fire-devastated mansion by a ruined orchard.

The images of death-resurrection cannot be denied. John Hardy suggests that Will is a likeness of Lazarus, newly risen from the tomb

(208), and the criticism is a fine one because Lazarus is a resurrection figure.[9] However, the allusions Percy makes to new life carry a deeper and abiding meaning of rebirth, namely that it is an unfinished process. While the resurrection of Lazarus was a great pre-figuration miracle, in Will's case, no miracle has happened and no miracle will happen. While his healing will continue in a ruined place, Will's recovery from the fall is slow, and although he learned something in the cave, nether he nor the reader has any sense of what it was.

"I was asking a question to which I resolved to find a yes-no answer."
"Did you find the answer?"
"Yes."
"Which was it?"
"I don't know."

Although Will is certain about something, the deliberate confusion regarding that certainty, the "yes", is as confusing as the question. Even though it can be surmised that Will went into the cave to discover if God exists, the persistent question could also be Will's haunting problem, "Are the Jews saved?" Even though Will Barrett desperately needs and is resolved to find the answer, he obviously does not know the question. His movement, as a wayfarer, is toward a knowledge and articulation of the precise question. When Allison has hoisted him to the table and washed him, she gazes down at him, and Percy describes what she sees:

The abdomen dropping away hollow under his ribs, the thin arms and legs with their heavy slack straps of muscle, cold as clay, reminded her of some paintings of Christ taken down from the crucifix, the white flesh gone blue with death. (SC 236)

Although critics have used the scene to characterize Allison as Mary Magdalene or the Pieta, in effect making Will the Christ figure, the analogy is confusing and cannot be maintained. What Percy is describing in the episode is death as the condition of all humankind, including Christ. The image of the body, cold as clay, has reverberations of the wages of original sin, "you will die." Will, lying on the table, resembling paintings of Christ taken down from the cross, is a wounded man,[10] like all men who are marked with the sign of death. Allison shares with Will the images of the crucifixion. She describes the electro-shock treatments as "Fried is crucified" when she refers to her death-like ordeal in the mental hospital. Graphically dramatizing the ancient and ever present spiritual condition of original sin, Percy shows Will and Allison on a journey of the wounded, both incorporating images of crucifixion, which is an experience of fixation to the cross. Obviously, the crucifixion images point to the difficulty of the human struggle, the cross each individual has been invited to carry.

But communitas is within Will's reach. He and Allison intend to marry and have a child (SC 344). Moreover, they will build garden houses on ten or twelve acre plots. Three old men from the convalescent home where Will lived for awhile will be employed in the project. Fulfillment which is found in participation is the joyous human consolation for the wayfarers on an otherwise weary journey.

At the beginning of *The Second Coming*, an alienated Will Barrett was unable to stand and to take command of his own posture, and at the end of the novel, besides the building of a residential village, he is forming a corporation, sharing the profits with Arnold and Ryan and other builders (346). Participation itself is essentially integration of the person into "communitas". Karol Wojtyla, (Pope John Paul II) says it well: "We are discovering the reality of participation as the quality of the self that acts 'jointly with others' basically because it finds its own fulfillment in this form of action" (42). Will Barrett is Adam, the fallen man in the fallen world who can articulate his infirmity as emptiness, loneliness, forgetfulness, and inability to stand. Although he is not always aware of the reality, a constant presence of Grace, God's promise, "I will never forget you," has been his salvation throughout his journey.

The marvelous journey of Will Barrett in *The Second Coming* is filled with splendid characters: Father Wetherbee, the old missionary whose faith is steadfastly associated with Apostolic Succession, Allison, whose language gets a bit mixed up so that she is "instigated by fear", the colorful Jack Curl, a Jimmy Swaggart of the modern novel. We recognize these people and we recognize ourselves as members of the same species, fallen individuals in a fallen world, confronting death in its many shapes and forms, and always consoled with the hope of rising.

Communitas for Will becomes his most valued vehicle of salvation. His willing participation with others in building and rebuilding a village, a place in the world to live with them, is essentially integration, the recovery of the self which otherwise is pitifully alienated. Communitas in Eden consisted of God conversing with Adam and Eve as the three walked in the quiet evening. Will Barrett, in his snatches of memory, evidently knows a perfection of community exists somewhere. At the end of the two novels about Will Barrett, the protagonist says "Am I crazy to want both, her and him? No, not want, must have" (SC 360). The HIM Will is referring to is old Father Wetherbee who represents "Apostolic Succession" for Will, the direct line to a redemptive journey in salvation history. Will, like Adam, wants to journey in the companionship of his love Allison, and the other member of the first community, God.

Marquette University

NOTES

[1] The term "uneasiness", which is often found in the novel and in Percy's essay "A Novel about the End of the World," very likely comes from Gabriel Marcel's *Problematic Man* in which the philosopher explores human uneasiness in twelve sections of his book, including "Uneasiness, Anxiety and Anguish", Uneasiness in the Perspective of the Gospel", and "Uneasiness in the World Today".

[2] Various scholars have opinions regarding the Peregrine falcon, like Martin Luschei [The Sovereign Wayfarer, 1972] who associates it with the central metaphor of "wayfarer, and William Rodney Allen [Walker Percy, *A Southern Wayfarer*, 1986] who claims that it represents Will's movement from a passive to an active man. In an interview I had with Percy on June 28, 1988, he told me that the bird was conceived in his mind by the poem "Windhover", by Gerard Manley Hopkins.

[3] In elaborating on the illnesses of both Will and Jamie, Frederick Crews maintains that Will's spiritual problem at the beginning of the novel is his undue reliance on destiny, on chance, and on unplanned interventions of fate. Jamie, on the other hand, is mentally sound but is drifting toward an ugly death that will show the engineer conclusively that he had better cease trusting in destiny (23).

[4] Percy's creation of Val fits into the paradigm explored by Cecil Eubanks in his essay "Eschatology and the Politics of Grace," in which he discusses revolution in its spiritual sense (132), viewing grace as a revolutionary possibility of change. Although Eubanks pays most attention to Binx Bolling and Will Barrett, Val most certainly is the figure who most fully echoes Isaiah the prophet (43:18), "Behold, I am doing a new thing." Her life has been revolutionized by Grace.

[5] The creation of the cave was "fun", Walker Percy said when he discussed it with Ben Forkner and J. Gerald Kennedy. The author claims there is a Lost Cove Cave in Tennessee and he just moved it to North Carolina. The warm air from the cave to maintain the greenhouse would be suitable for the idea Percy was working with, namely, the starting of new life.

[6] Percy himself is not a stranger to waiting, as he claims in a discussion with his friend, Robert Coles: "I think I once wrote to you that I sit and wait, as Kafka told himself to do. I look out the window. I stare at the ceiling. Then I write" (240).

[7] In the Epilogue of his book on Percy, Robert Coles remarks that the notion of ruin is inevitably present in the author's work although Percy approaches the topic with a sense of humor and a mixture of amusement and anxiety. Coles notes that a certain attitude exists for the author, a "studied distance from the temptation to self-satisfaction; and apprehension that the whole show [twentieth century progress, much acclaimed] may soon

enough collapse". The "ruins" of his last two novels venture to anticipate the distinct possibility of collapse (239).

8 Patricia Poteat is on target when she says that for Allie, rebirth is to a new life she finds in lanugage – not in the theories, but with the ordinary business of speaking (166). She has had the problem with the silence that fell between words, and wondered which was worse, silence or speech. Will, however, can interpret her language, and thus she is reborn with his help.

9 Examining the use of the death motif with its relation to the Christian imagination, William Lynch says " . . . human temporality can be discerned both by the honest literary imagination and by the theological or Christological intelligence to be moving through many forms of death. It seems to take death as its weapon and by some sort of extraordinary process, is always rising through it to insight in the natural order and resurrection in the supernatural order" (31).

10 Describing the unconscious Will, Percy uses medical terms which Robert Coles insists are part and parcel of his writing. "He has been for thirty five years, still is, and will be to the last breath a doctor" (239). Coles notes that Percy has the physician's familiarity with the body and with death.

REFERENCES

Brooks, Cleanth. "Walker Percy and Modern Gnosticism" in *Walker Percy*. Edit. Harold Bloom (New York: Chelsea House Publishers, 1986).

Coles, Robert. *Walker Percy: An American Search* (Boston: Little, Brown and Co. 1978).

Crews, Frederick C. "The Hero as 'Case'; The Last Gentleman" in *Walker Percy*. Edit. Harold Bloom (New Haven: Chelsea House Publishers, 1986).

Forkner, Ben and Gerald J. Kennedy. "An Interview with Walker Percy – 1981" in *Conversation with Walker Percy*. Edit. Lewis Lawson and Victor Kramer (Jackson: University Press of Mississippi, 1985).

Hardy, John Edward. "Percy and Place" in *Walker Percy: Art and Ethics*. Edit. Jac Tharpe (Jackson: University Press of Mississippi, 1980).

Lynch, William, S. J. *Christ and Apollo: The Dimension of the Library Imagination* (New York: Sheed and Ward, 1960).

Marcel, Gabriel. *Problematic Man*. Trans. Brian Thompson (New York: Herder and Herder, 1967).

Percy Walker. Personal Interview, 28 June, 1988.

Percy, Walker. *The Last Gentleman* (New York: Farrar, Straus and Giroux, 1954).

Percy, Walker. *Message in the Bottle* (New York: Farrar, Straus and Giroux, 1954).

Percy Walker. *The Second Coming* (New York: Farrar, Straus and Giroux, 1984).

Poteat, Patricia. *Walker Percy and the Old Modern Age* (Baton Rouge: Louisiana State University Press, 1985).

POL VANDEVELDE

THE POSSIBILITY OF A PHENOMENOLOGY OF THE TEXT:
From and Against Postmodernism

It is a common theme in Postmodernism that a "text" or a "discourse" creates its own object, that, as Derrida says, "there is no outside of the text" or that "everything is interpretation". Usually, Postmodernist thinkers call upon German Romanticism, especially in the person of Friedrich Schlegel, upon his role in the delineation of our modern concept of literature and also his influence on philosophy and literary criticism. In using the concept of allegory, as opposed to symbol, and in applying these concepts to the theory of literature, it is widely accepted that Schlegel brings to its end a conception of literature based on the mimesis, i.e. on the transitivity of signs referring to the external world. Schlegel is credited with the claim, by and large, that literature is no longer or not only allegorical, but essentially "symbolic": literature has no transitivity and refers only to itself. That means that, the concept of "literature" signifies a production – and no longer a condition of scholars or a quality of educated people – it also means its own theory, i.e. that literature has reflection upon itself. Literature presents itself in producing its own theory. One can quote Fragment 115 of the *Lyceum*, a year before the foundation of the *Athenäeum* review by the Schlegels, which enunciates the program of this review: "All art must become science and all science must become art; poetry and philosophy must be united".[1] In the *Athenäum* review, we can read that "All poetry is poetry of poetry; for it already presupposes language, whose invention pertains to poetic activity, and which is itself a poem of humankind" (KA II, Frag. 238). To this extent, such is the general picture, a literary text has no external object, but has only an object insofar as it is grasped, identified, or described through the text. A text, thus, properly "creates" its own object.

In France, a lot of critics, especially in what is called *La Nouvelle Critique* (Roland Barthes), and many philosophers, the ones adhering to so-called Postmodernism or Post structuralism, have taken up this view and analyzed the change in the nature of literature. Philippe Lacoue-Labarthe and Jean-Luc Nancy have written a beautiful essay on this

M. Kronegger and A-T. Tymieniecka (eds.), Analecta Husserliana XLII, pp. 277–291.

subject: *L'absolu littéraire. Théorie de la littérature dans le romantisme allemand.* Michel Foucault, for his part, has analyzed the shift at the end of the 18th century, where, according to him, the literary act becomes aware of itself and ceases to be the echo of a first voice. It drives itself toward the pure and inaccessible essence of literature. The literary act becomes a transgression. "One makes sound the infinite murmuring, the piling up of words which have already been uttered".[2] Literature is literature in designating itself, in taking place within a universe of discourse from which literature speaks as the one which speaks of itself. For Foucault, this is the acknowledgment that there is no outside for literature.

It is also a common theme in contemporary philosophy or literary criticism that so-called Postmodernism or Poststructuralism originated with Martin Heidegger. Usually, Postmodernism includes thinkers like Michel Foucault, Jacques Derrida, and Jean-François Lyotard, among many others.

As a matter of fact, Martin Heidegger offers a view on literature and philosophy (or theory in general) that appears revolutionary in two senses. Firstly, it is a continuation and accomplishment of several programs sketched by romanticism: the work as its own theory, a universal poetry (Schlegel); language as the spirit of a people and, to a certain extent, a masterpiece (Wilhelm von Humboldt). Secondly, Heidegger has exerted a heavy influence upon, and prepared the way for, the French "Nouvelle Critique", deconstructionism, and "Postmodernism".

Heidegger is concerned with the self-consistency of the artwork. He accounts for the failure of the traditional aesthetics which is exhausted at the moment at which the work turns away from the paradigm of "mimesis". If the work escapes a mimetic task, it asserts, thus, a certain "intransitivity"; that is, it is not a means for something else, nor a mediation to something else. In going a step further, we can say that this self-reference in the case of literature means that the opacity of the work is no longer hermeneutic. In other words, the opacity does not concern itself with what is to be represented, to be imitated, or to be translated. What we call "the world" consists only in the text itself, in its signifying structure, and its "machinery" for producing sense, as Roland Barthes put it. The self-reference of literature amounts to a self-designation. In a similar way, this self-reference is a self-assertion: literature does not imitate literature – as an aesthetic gesture – but speaks about literature, or, as we say in French, "parle la littérature" (speaks literature), i.e. shows itself as literature, in a manner which is not synonymous with repetition.

All of these themes and topics are fascinating. They constitute a trait of our Modernity. It is from within this network of re-interpretations that the methodological terms of "reading", "writing", "deconstruction", and "archaeology" can make sense. The "linguistic turn" brought about the view that one cannot think without concepts or talk without words. The renewal of interest in literature and its functioning goes a step further in assuming that one cannot think or talk except about what has already been thought or talked about.

However, without denying the importance and relevance of the present theory of literature or philosophy of literature, I will try to show that the role of Romanticism has been overemphasized in the present theory of the text. I would like to examine, first, whether this general picture of the romantic literature as its own discourse holds true and passes the test of a reading of several texts by Friedrich Schlegel; secondly, I will investigate to what extent Heidegger, considered the forerunner of Postmodernism, is a contemporary heir of romanticism and whether the conception of the "text" as creating its own object can be found in his works. I will spell out three themes and try to find out the treatment they received, first, in Fr. Schlegel, and, second, in Heidegger. These three themes are the following: 1) poetry and intransitivity, 2) poetry and interpretation, and 3) poetry and theory.

1. FRIEDRICH SCHLEGEL'S UNIVERSAL POETRY

In reading Friedrich Schlegel's "Fragments" for the review *Athenäum*, his *Gespräch über die Poesie* of 1800, and his *Philosophische Vorselungen insbesondere über Philosophie der Sprache und des Wortes* of 1828, what is the real meaning and impact to be found for these three themes? What does it mean, first, that poetry is characterized by a radical intransitivity?

1) *The Intransitivity of Poetry*

What does it mean that there is only a poetry which is the poetry of poetry? Or what does it mean that "the poet's arbitrariness (*Willkür*) does not suffer any rule to be imposed upon itself" (Frag. 116, *Athenäum*)? Fragment 116 in the *Athenäum* begins with the famous motto: "Romantic poetry is a progressive universal poetry" (KA II, p. 182). As explained further by Schlegel, "progressive" means that "the romantic way of poetizing is still in becoming; it is even its authentic being that it can only be in becoming for ever, that it can never be completed" (Frag. 116). Criticism or a theory cannot thus characterize completely such a poetry. In this respect, "Poetry can only be criticized by poetry" (*Lyceum*, Frag. 117; KA II, p. 162). If "Poetry and philosophy should be united" (*Lyceum*, Frag. 115; KA II, p. 161), it is not through the unification of two pre-existing genres, but the task of Modernity to find its specific way and mode of expression. Schlegel uses the expression "poetizing philosopher" as well as "philosophizing poet" to sketch this modern necessity. Such a poet or a philosopher (and it should be the same "person") is a prophet (*Athenäum*, Frag. 249; KA II, p. 207).

Universal poetry thus means a new art of expression which includes the traditional disciplines of philosophy and poetry, but also the sciences; and the reason is not, as pointed out by Postmodernism, the status of the signs, or writing general, but, for Schlegel, the very link with the outside world: there is a perfect harmony between the poetry in the arts (*Kunstpoesie*) and the poetry of nature (*Naturpoesie*). There is "a shapeless and unconscious poetry which is active in plants, shines in the light, smiles in a child" (*Gespräch über die Poesie*, KA, II 285). This "nature poetry" is "the first and original poetry, without which there would be no poetry of words" (*Gespräch* . . . , KA II, 285). As humans we ourselves are "the poem of the Divinity" (KA II, 285).

From these quotes it appears that poetry is not a genre, is even not primarily made of words, but is more an instinct. Wilhelm von Humboldt used the expression "Sprachkraft" to name a kind of linguistic instinct. In the same way, Schlegel wrote that poetry is possible thanks to an "elemental force of humanity" ("eine Urkraft der Menschheit", KA II, 285). When he writes that it is only in poetry that one can speak about poetry, what is at stake is not so much self-reference, but rather an ability to be poetic; poetry is a force, not a writing. Only in poetizing may one be, not only a poet, but a human being. It is in this sense that romantic poetry as universal poetry is a philosophical poetry. Schlegel can invert the formula and write that it is a poetic philosophy as well; and this poetry of poetry can be articulated in a philosophy of poetry or a philosophy of the novel.[3]

2) *Poetry as the Use of Symbols*

Up to this point we have a picture of Schlegel which is somewhat distant from that of a postmodernist forerunner. We mentioned in the beginning the revolution with which Schlegel has been credited in his opposing symbol and allegory in his putting them on a thoeretical level: to have made literature a symbolic activity, designating itself and creating its own universe, against openness to the world, or the mimesis typical of an allegory. In *Gespräch über die Poesie*, we can read: "All beauty is allegory. Because it is inexpressible, the Highest (*das Höchste*) can only be said allegorically" (KA II, p. 324). What is interesting for us is that another version says: "the Highest can only be said symbolically".

I do not mean that there is an ambiguity in Schlegel. After all, there are only ambiguities in Schlegel. However, the ambivalence or hesitation between allegory and symbol leads us to put into question the rigid distinction drawn by Postmodernist thinkers in the development of literature. The opposition between symbol and allegory is also rather purely methodological and heuristic.

I see at least two reasons why Schlegel himself hesitates between poetry that is a traditional mimesis of life and the world, on the one hand, and poetry as the productive force of the interpretation of the world, on the other; i.e. between pre-modern and post-modern views, if I may say so. The first reason is his idealism: he believes that the mind, the spirit, gives its rule to the world, and does not discover an already-laid-down and organized world; however, this mind is not only and not strictly "human". The force present in humans under the form of the mind is a gift of the Divinity; the mind itself is the spirit of the universe. Schlegel is idealist in the sense of a universal idealism: the spirit is everywhere. The second reason is tied up in his conception of language: "language or telling which is [. . .] identical with the symbolic force, is the first and immediate tool through which spirits work and come into magical contact with one another" (KA II, p. 354).

The ambivalence between allegory and symbol makes poetry a double-sided figure, the Janus figure of literature right up to now. Poetry is made of signs, and to this extent has a referential character. Literature "names" something, and this something does not pertain to the order of signs. However, this general ability to name as well, as the impossibility of relating to the world otherwise than through signs, makes literature a productive activity. Expressions, the uses of words, qualifications, ideals, ways of characterizing things, feelings, modes of understanding human relationships are all described and moulded by literature.

All this leads Schlegel to affirm that every art is and can only be "poetry", even though it is not made of words. It is poetry in a double sense: it is a poetizing, a making, a producing; and, secondly, it has a link, even though not obvious, with language, which is the immediate way, and the only one, through which the world may be understood and shaped.

3) *Theory as a Part of Poetry*

In the light of the previous remarks, we can now understand more accurately what Schlegel means by a theory which is a part of poetry. From this point I will treat his lectures entitled *Philosophische Vorselungen insbesondere über Philosophie der Sprache und des Wortes* given in Dresden in December 1828.[4] In these lectures, Friedrich Schlegel stresses the creative power and task of criticism: "Criticism must not judge works according to a general ideal, but find out the individual ideal of every work"; "criticism compares a work with its proper ideal".[5] He also states that "the science of art is its history" (KA II, p. 290). There is no atemporal poetry, there are no eternal rules; the science of poetry is merely a history of the processes of doing and making poetry.

Schlegel is also the first to mention and develop the "hermeneutical circle". A critic must start from the word, then go to the sentence, then to the whole work, and, lastly, come back to the word. Schlegel also makes a distinction between the meaning of a work, the "real individual meaning of an author", and the "logic and grammatical" question of the meaning of a text – in other words: between the meaning desired and intended by the author, and the meaning apprehended by the reader in his/her interpretation.

In this context, Schlegel says that "every work [. . .] knows more than it says and desires more than it knows" (KA II, p. 140). That is why the task of criticism is that of helping the work to reach its full development, to unfold itself in its full sense. Criticism "through its own creativity has the task of completing a work" (KA XVIII, p. 48, Frag. 299). "To criticize somebody means to determine his/her fragments and projects" (KA XVIII, p. 48, Frag. 299). To this extent, criticism is also and essentially "productive". Schlegel formulates the motto of the modern hermeneutics proceeding from Schleiermacher: "To criticize means to understand an author better than he/she understood him/herself" (KA XVI, p. 168, Frag. 952).

However, this task of criticism, that of completing a work of art, given its fragmentary character, only means for Schlegel that a "Geist", a spirit (what he also calls "chaos", or "system") pervades it. The aim of a critic is to help the work reach its full development, but that means to manifest and reveal the universe already present in the fragmentary work, or to complete the unfinished work which bears its principle in itself.

What is striking in these lectures is the constant emphasis on the "Geist", the spirit behind and prior to the literary work. There is no "self-designation" in a work other than the fact that a literary work manifests the spirit of humankind. Nowhere can one find in these lectures anything that even sounds like a "resistance to theory" (to use Paul de Man's formula). Nowhere can one find the impossibility of going behind the semiotic texture of a work. For Schlegel, there is an outside of a text; a text is even "outside" only: there is a "Kunstpoesie" because there is, prior to it, a "Naturpoesie".

Never does Schlegel claim that criticism has a productive power over the principle of the work itself, nor does he believe that this principle is an object of discussion or of interpretation. Nowhere can we find a place where Schlegel says that a work only exists in its interpretation or that a work has no outside. On the contrary, if a work needs criticism, it is only because it is overwhelmed by its outside. A work, for Schlegel, has only a meaning in relation to what the work aims at: the universe, the chaos, the system, the "Geist". A work has the privilege of being the bearer of the voice of the spirit and must have the humility to be only the support for the realization of its real spirit. Schlegel qualifies the original language which is concealed in art as "eine innere geistige Ursprache". True art and great poetry awaken echoes of the "Ursprache". They are echoes of the soul.

2. HEIDEGGER: A WORK OF ART AS DISCOURSE

With regard to Heidegger, I will only focus on the relation of discourse or language to a work, as presented in the lecture "The Origin of the Work of Art".[6] Heidegger's presentation can be summed up in the following theses: 1) a work is intransitive, 2) a work unfolds itself within an interpretation, 3) a work is "Dichtung" (poetry).

1) *A Work is Intransitive*

A work can only display itself and function as a work between the two moments of its creation and reception. Heidegger uses many formulations to deny any form of mediation in a work. A work does not say anything, does not communicate anything, does not represent anything. A work is not an allegory, nor a symbol, nor an expression. If art is expression, it is, Heidegger writes in the first version of "The Origin of the Work of Art" (1935), because it is art. If that be not the case, "The Acropolis is the expression of the Greeks, and the cathedral of Naumburg is the expression of the Germans, and bleating is the expression of the sheep".

A consequence of this intransitivity is that a work is not only the result of a making, but more primarily an event. A work is an event characterized by its punctuality and is related to time. Before its being-there, Heidegger says, a work was not yet, and it will not be any more after the work. The arising of a work is a thrust into the world wherein it appears. The works we know are in the mode of "having-been". The works "will never be what they were" (Bd. 5, p. 26/41); "a work belongs as a work to the realm it opens itself" (Bd. 5, p. 27/41).

2) *A Work Unfolds within an Interpretation*

If a work is linked to the punctuality of its appearance and arising, it is not really itself thereafter. Its being extends itself and continues in another dimension. This dimension wherein a work extends itself and unfolds, is what Heidegger calls its "preserving". A work is only a work for its "guardians", i.e. for "a historical people" (Bd. 5, p. 63/75). A work finds an echo in a preserving which is characterized as a "knowledge" (Bd. 5, p. 55/70), in other words "philosophy". A work is created by an author, but this creation finds the full resonance of its arising only in a preserving that permits it to "speak". "Only such knowledge prepares its place for art, their way for the creators, their location for the preservers" (Bd. 5, p. 66/78). Creation needs such a preserving. There is a solidarity between creation and preserving; Heidegger calls it a "creative preserving" (*schaffende Bewahrung*). The creation of a work has a meaning only in the event of the arising of the work. What an author intended to do lies beside the work as an event, and the only access we have to the creation is through the work. The transitive path from the author to the work is lost, or it is a myth. Only a reconstruction is possible. However, be this reconstruction as disinterested and faithful as it may, it will never recover the act of creation in its singularity and uniqueness, nor the work in its might. The author him/herself has no privileged access to his/her works, but can only interpret.

According to Heidegger's first thesis, a work unfolds and displays itself within an interpretation. A work needs a preserving in order to be what it is. But this preserving does not aim only at disclosing what a work would not say or show clearly. The solidarity between both creation and preserving means that neither of them can stand alone. Creation cannot claim to give an account of a work in its event-character, nor can preserving obliterate the moment of creation. To interpret does not mean that the interpreter can jump into the reference which a work bears in making clear and laying down what the work means. If a work needs a discourse that speaks after it and from it, this discourse that interpretation or preserving is must unfold itself back to the arising and the event of the work. The work itself is the interpretation. The work itself is a discourse. This discourse will only speak in the discourse held and given on it. The duplication aims precisely at preventing an infinite referring of discourses where texts mean other texts ad infinitum.

A work for Heidegger is what I call an "interpreting interpretation": what a work says, that is, its own origin. On the one hand, a work modifies our relation to beings in general by impressing on us our distance from these beings; a work makes the relation we have to the world present. A work shows this relation to us and reminds us that it is a relation, while we generally hold it to be usual and, thus, natural. In Heidegger's words, the temple of Zeus at Paestum was for the Greeks an interpreting interpretation in that social life was organized around the temple, in a certain conception of the divinity and in the spirit of the sacred. People who live in the neighborhood and under the protection of the temple are not aware of that. They were not unaware either, but they showed in their life the presence of the temple and their understanding of it. The temple could be constructed and be the space where a God is celebrated only because the temple was borne by an interpretation (an interpretation of what life is, of what death, a human, his/her role in proximity to God, etc., are). This interpretation is not an interpretation given to the temple. On the contrary: the temple has been built on the basis of this interpretation. The rockbottom foundation of the temple is an interpretation.

As an interpretation, the work does not simply appeal to what it depicts, nor to any immediate interpretation of it as artwork. Instead, the work is only able to be an interpretation in its pointing to itself as an interpretation of. The temple at Paestum is not only a place of worship. The temple itself makes the place a place of worship and the temple itself makes God present in the worshipping which the temple circumscribes. The temple lying upon an interpretation is itself an interpreting – not that it interprets or shows God; rather it allows God to be present.

There is, thus, a passive and active character in a work of art. It is passive in the sense that a work lies upon an interpretation of beings taken for granted and which organizes, as if a principle, the relations between things themselves and humans. Michel Foucault used the term "orders of reality", which he called the "episteme". The active side of a work is the fact that a work re-configures and re-shapes the relation to beings that held true up to then in making this relation appear as such.

We may sum up what "interpreting interpretation" means as follows:

1. A work lies upon an interpretation; it does not originate from nothing. A creation brought it to its form.
2. A work is not a decorum of a culture or a cultural feature of a community, nor is it an expression of this culture. A work is an interpreting in its constituting the glue of a culture, in its shaping of an organization of or an order of things and of humans related to things.
3. Since this interpreting by a work does not involve any transitivity, since a work has no private language that it could unfold and display for humans, this activity only receives its weight, its meaning and its strength from people who understand the work as a re-configuring of their relation to beings. Only due to the responsibility of "preservers" can a work be properly a creation, in the sense of an event in history. A work is an interpreting only in the interpretation which finds its legitimation in it and brings the work to its shining forth, to its "speaking". The restrained "speaking" of a work speaks in the way of a "having-been": "Dieses hat gesprochen", Heidegger says. The work spoke, not in the past, as in a speaking which was once present, hearable and recognizable. Rather, the work spoke "in the present", because it needs to be heard in order to be interpreted; and this interpretation only becomes a "speaking" through the responsibility of the listener. As the lecture "Hölderlin und das Wesen der

Dichtung" (1936) ("Hölderlin and the Essence of Poetry") makes clear, interpreting only becomes "wordly" (*worthaft*) as responsibility (*Verantwortung*). In other words, what someone says in interpreting a work will never be legitimated by the work, since the work stands alone and does not say anything. Only in the "having-been" of the work of art does the interpreter's discourse find its echo, as the resounding of what cannot sound.

4. That means, then, that the active and creative side of a work only displays itself within the framework of art. The Temple at Paestum only shows that of which it is made, its matter. It shows itself as the one which shows. While showing itself, a work projects itself anew. Insofar as the work displays itself, it also projects what holds true of a work, and reshapes the work-being of the work each time it appears as such. This is an exposition and a reformulation of what the essence and the meaning of art is (the "essence" as what is recognized as art at a time, in a culture; and the "meaning" insofar as this "essence" is related to history, as that which has its roots in such a history as the history of interpretations). It re-configures again the essence of a work and determines anew what art is. In this general and comprehensive sense, art is "the setting-into-work of the truth", the reshaping (*Gestalt*) of the truth and the origin of the truth.

3) *A Work of Art is Poetry*

The two features of a work, its being intransitive and its being an interpretation, are summarized in the qualification of a work as "poetry" (*Dichtung*). The intransitivity we mentioned appears to be rather a self-reference. A work only designates itself and "speaks" of itself. Hölderlin, for example, is qualified as "the poet of the poets". However, this self-reference is not the fact that literature, for instance, speaks about literature, or that a work creates its own object. In other words, if it is clear that for Heidegger a work is not an allegory in the sense that a work does not directly name or describe "the" world; his position does not amount to making the work a symbol (in its technical opposition to allegory). A work is not enclosed in itself. The self-reference Heidegger evokes is not a mere duplication according to which poetry speaks about poetry. Such a mere duplication is, for Heidegger, a "highly modern reflection", where a "sickly reflection spreads into poetry (in such a way) that it always comes around upon itself".[7] Hölderlin is not the one who repeats, but the poet who takes up poetry in this original unfolding and, thus, experiences its might; only in grounding poetry anew, "does he project it again for an epoch".[8]

To this extent, every work of art is a "Dichtung". Such a claim assumes that art is made of language. What does this mean? If a work is only qualified as such in and through an interpretation, this entails that a work – even a painting which is not in itself linked to language or made of language – necessarily pertains to the order of language. A work is in the potentiality of language, in the potentiality of being spoken of or talked about. This potentiality allows a work to be interpreted. Language is thus the medium which is common to both the work and the interpreter. Language is the medium in which a work can reach its presence, and language is the medium through which the interpreter articulates his/her understanding of it. Complementarily, this also means that a work bears and promotes the word spoken from it.

In this sense, a work may be qualified as a discourse. It is not that a painting, for instance, can be or must be "translated" or "transposed" into language. Heidegger incessantly repeats that a work does not say anything, does not communicate anything, and does not express anything. This means also that a work is not the form taken by a previous discourse, as though, before the specific shaping of a work, an author's intention or a social organization served the function of a discourse. If a work is a discourse, it is not in the sense that discourse is the internal speech of the work, its meaning-kernel. The discourse that a work is consists in the interpretation a work is, through which it asks to be interpreted.

Heidegger summarizes all these features under the name "Dichtung" (poetry). The use of the word "Dichtung" is interesting because it ties a literary, or "written" activity to such arts as painting or sculpture. That aleady gives a hint about the role taken by language. "Building and plastic creation [. . .] always happen already, and happen only, in the open of saying and naming [. . .]. They are an ever special poetizing within the clearing of what is, which has already happened unnoticed in language" (Bd. 5, p. 62/74). The temple of Zeus at Paestum constitutes a fable, a legend, a naming (*Nennen*). While it is still vague, we can see that "such saying (*Sprache*) is a projecting of the clearing, in which announcement is made of what it is that beings come into the Open 'as' (. . .)". Projecting saying is poetry (Bd. 5, p. 61/73–74).

A work is "Dichtung" (poetry) in the double sense of "dictare" and "poiesis". A work is a making, a creation, a production which happens between the act of creation and the preserving. A work is an event. The poetic moment in the work constitutes what we call the resistance of the work to interpretation, and renders the interpretation as a kind of "reading", in the sense Paul de Man would understand. However – and this is the main difference with a deconstructionist or postmodernist view – a work is also, for Heidegger, a "dictare". The word *Dichtung* shares roots with the Latin *dictare*, from *dicere*, to say, which is derived from a root common to the Greek *deiknumi*, to show. A work is a saying, not only because of its own power, but primordially because it gives words to its origin. A work is such in the destiny within history, wherein creation and preserving are inscribed. This means that ultimately all the levels of interpretation we mentioned (the interpretation upon which a work lies, the interpretation proper to preserving wherein a work "speaks", the interpretation that the work is as interpreting interpretation, and eventually, the interpretation of what a work and of what art is) have themselves a provenance and have been themselves projected. They only shape the texture of a more original saying, constituting what Heidegger calls history-ladenness (*Geschichtlichkeit*). For Heidegger, this history-ladenness of a work covers the history of metaphysics, which is for him the succession of different interpretations of beings.

"Dichtung" names a mythical moment of the becoming-language, becoming-art, becoming-poetry. The point here is that, if we are now moving in a world of discourse, if we are the story tellers and the speakers, then we are such only in a history. This history is that of the tension between story and events, between words and things, between fiction and reality. What Heidegger intends is not to regain or to recover the pre-Greek moment where the word and the thing were one, but rather to show the way to return. Such a return should take its point of departure in our actual situation, wherein we have to tell the story of this opposition which has led us to being the ones who are between fiction and reality.

CONCLUSION: FROM DISCOURSE TO TEXT

We saw that Schlegel is not the forerunner of our concept of the text. Poetry, even the famous program of a universal poetry, only works as a multifaceted mirror, and what can be mirrored is our "elemental force", that which makes us the human beings we are, and which is called "poetry". Common to Schlegel and Heidegger is at least this deep belief and conviction of a principle in the universe. Schlegel refers to it as a Spirit. For Heidegger, it is Being, unfolding itself within the History of Western thinking, something like an ultimate transcendental measure. This is their answer to Hölderlin's question: "Gibt es auf dieser Erde ein Mass?" For both Schlegel and Heidegger, literature, or poetry, is not made of "texts".

Heidegger must be granted to have given a subtle articulation to vague romantic views on poetry and language. In this regard he escapes the reproach of paradox, to which many romantic writers are inclined: Schlegel on literature or Humboldt on language.

Even though Heidegger understands poetry or literature as discourse, there are no postmodern accents in his views. Heidegger may repeat that a work pertains to the order of language, is made of signs. He holds, however, the firm belief that these signs have an archi-sign: the truth as the destiny of history in *The Origin of the Work of Art*, or as "the voice of the Gods" in his first lecture course on Hölderlin in 1934. The semiotic function of a work is paralyzed as soon as it is made of signs. The signs can only refer to their origin, the archi-sign, or a transcendental signified, Being. All this means that, even against what Heidegger overtly claims, another semiosis intervenes which functions without signs, that which is the understanding of the showing forth of signs, that which is the "making" of "poiesis" and the dictating of "dictare". This other semiosis is thinking itself, which saves and preserves the voice of the poet from the flux of human words.

A text, henceforth, cannot and will never be a text for Heidegger. It has an existence only in so far as knowledge uses it in its movement to interpret in order to attain through the work the destiny of history. As stated in the course on Hölderlin in 1934, a text is a semblance (*Schein*).[9] The saying of Being, Heidegger says, is not first present in libraries, bookstores, and print shops. Before all that, this saying permeates the language of a people. As soon as a poem is a text, for Heidegger, it is only present and available (*vorhanden*), the object of study and of talkative concern. It sinks into communication. It is the task of thinking to render a text transparent in order to "see" the true poetry *in actu*.

Schlegel was at the origin of this view according to which a self-contained and self-sufficient text is impossible: between Spirit and Letter

(*Geist und Buchstaben*); Schlegel's goal is "to fight against the vanishing of the Spirit in the letters which in themselves and taken for themselves do not say anything".

NOTES

[1] Friedrich Schlegel, *Kritische Ausgabe*, Bd. II, "Charakteristiken und Kritiken I" (1796–1801), ed. Hans Eichner (München, Paderborn, Wien: F. Schönig Verlag; Zürich: Thomas Verlag, 1967). I will use the abbreviation KA for *Kritische Ausgabe*, followed by the volume and the page or the number of the fragment. The translations are my own.

[2] "On fait entendre l'infini murmure, l'amoncellement des paroles déjà dites", in Michel Foucault, "Langage et littérature", lecture given in Brussels, manuscript.

[3] I leave aside the question of the meaning of "Roman" in Schlegel, linked as it is with "romantisch", "romanisch", and the narrative form of the novel. Cf. the introduction by Hans Eichner to *Kritische Ausgabe*, II.

[4] "Philosophische Vorlesungen insbesondere über Philosophie der Sprache und des Wortes", in *Kritische Ausgabe* X, ed. E. Behler (München, Paderborn, Wien: F. Schönig Verlag. Zürich: Thomas Verlag, 1969.) The translations are my own.

[5] "Von der Schönheit in der Dichtkunst", in *Neue philosophische Schriften*, ed. Josef Köner (Frankfurt a. M., 1935), pp. 363–387.

[6] M. Heidegger, "Der Ursprung des Kunstwerkes", in *Holzwege* (1935–1946), *Gesamtausgabe*, Band 5 (Frankfurt a. M.: Klostermann, 1977). "The Origin of the Work of Art", in *Poetry, Language and Thought*, trans. A. Hofstadter (New York: Harper & Row, 1971), hereafter "GA," followed by the volume. The pages are given in parenthesis, first, the German text, then, the English translation. In my quotes I slightly modify the translation. This lecture was given for the first time in Frankfurt a. M. in 1936. Now we know that there were two other versions; the first was only a draft (1935) and is now available in the review *Heidegger Studies*, in a transcription by Hermann Heidegger; the second version is a lecture given in Freiburg in 1935. The text of this lecture was recently published by E. Martineau with a French translation. We cannot take into account here the deep rewriting and change of perspective between the two first versions and the last one of 1936, with which we are here essentially concerned.

[7] Martin Heidegger, *Hölderlins Hymnen "Germanien" und "Der Rhein"*, ed. S. Ziegler, GA, Bd. 39 (Frankfurt a. M.: Klostermann, 1980), p. 218. My own translation.

[8] *Ibid.*, p. 219. My own translation.

[9] *Ibid.*, p. 256.

DAVID L. MOSLEY

GUSTAV MAHLER'S *ICH BIN DER WELT ABHANDEN GEKOMMEN* AS SONG AND SYMPHONIC MOVEMENT: ABDUCTION, OVER-CODING, AND CATACHRESIS

While many have commented upon the manner in which Gustav Mahler quotes both himself and other sources in his music, this stylistic feature of Mahler's composition has not, to my knowledge, been examined from the perspective of intertextuality. The many intertexts of Mahler's music include: literary texts of his own creation as well as those by other authors, other musical styles and genres, musical compositions by other composers, and portions of his own oeuvre. Nowhere is Mahler's musical intertextuality more informative than in the many and varied relationships between his songs and his symphonic compositions. The case of *Ich bin der Welt abhanden gekommen* as poem, piano-Lied, orchestral song, and the Adagietto movement of his Symphony No. 5 is one such example. This essay will employ a semiotic methodology to examine: (1) the *abductive* process by which Mahler approached Rückert's poem, (2) the practice of over-coding by which this poem became a piano-Lied, an orchestral song, and finally a purely instrumental symphonic movement, and (3) the manner in which the Adagietto of Mahler's Symphony No. 5 constitutes a *catachresis*, or misapplied metaphor, in relation to the text from which it originated.

MAHLER'S ABDUCTIVE APPROACH TO LIED COMPOSITION

What does it mean for a composer at the beginning of the twentieth century to quote the musical material from a vocal composition in a subsequent instrumental composition? The first implication of this question has to do with our model for understanding Lied composition itself. The growing independence and autonomy of the musical material for a vocal composition, which is exhibited throughout the nineteenth century, is best understood as an evolution from an *inductive* attitude on the part of he composer toward the text to an attitude which is best described as *abductive*.

Charles Sanders Peirce characterized abduction as "a method of forming a general prediction" (2.270). It "consists in studying facts and

M. Kronegger and A-T. Tymieniecka (eds.), Analecta Husserliana XLII, pp. 293–301.

devising a theory to explain them" (5.145). For Peirce, the distinction between induction and abduction was not always easy to make.[1] He admits, "in almost everything [he] printed before the beginning of this century . . . [he] more or less mixed up Hypothesis [Abduction] and Induction" (8.227). An abductive inference carries no guarantee of certainty, but "merely suggests that something *may be*" (2.96). While Peirce seems to limit his application of induction and abduction to scientific inquiry and the constraints of propositional logic, these modes of thought also have important aesthetic implications and applications.

Peirce's understanding of induction is conventional, i.e. it involves the observation of a sufficient number of individual facts, and, on the grounds of *analogy*, extending what is true of these facts to the observation of other facts of the same class, thus arriving at general principles or laws. Abduction, however, begins with the inference, or hypothesis, that certain facts or observations might be explained by such and such a theory and then proceeds to explain those facts or observations *as if* the inference, or hypothesis, were true. Induction, then, is a comparative process based upon *analogy*, while abduction is a creative process based upon *metaphor*.

When applied to the matter of song composition, the concept of induction describes the manner in which Lied composers of the late-eighteenth and nineteenth centuries set a poem to music. Composers like Reichardt and Zelter took it as their task to mirror the formal elements of a poem in music. This approach posed little threat to the linguistic meaning of the poem and may explain why Goethe, for instance, valued the imitative nature of Zelter's settings of his poems (Hecker, 9). Schubert and Schumann were less interested in mechanical imitation and, instead, seized upon certain semantic elements of the poem for which tonal analogues could be found. Through the systematic use of these tonal analogues with each successive reference to the semantic element, their settings created a correspondence between words and tones.[2] This approach involved certain choices on the part of the composer in reference to the linguistic meaning of the poem. Lawrence Kramer has referred to these choices as constituting a "refined form of erasure" in which the setting obliterates certain semantic possibilities and emphasizes others (129). This may explain why Goethe was less kindly disposed toward Schubert's settings of his poetry. Schumann summarizes this approach to Lied composition when he makes the following statement in reference to Norbert Burgmüller's Op. 10: "He regards it – as everybody should – as the highest task to recapture the poem in its smallest detail with finer musical material" [*Das Gedicht mit seinen kleinsten Zügen im feineren musikalischen Stoffe nachzuwirken, gilt ihm das Höchste, wie es Allen gelten sollte*] (175). And again, in praise of his contemporary Robert Franz, Schumann states, "he strives to represent an embodiment of the poem" [*er will uns das Gedicht in seiner leibhaftigen Tiefe wiedergeben*] (348). Gustav Mahler's approach to song composition involes the positing of a single musical construct which is then developed within the code of music as a musical hypothesis about the meaning of the poem.

Mahler describes the process by which he sets a poem to music in the following manner: "you can express so much more in music than the words directly say. The text is actually a mere indication of the deeper significance to be extracted form it, of hidden treasure within" [*Man kann mit der Musik doch viel mehr ausdrücken, als die Worte unmittelbar sagen . . . Der Text bildet eigentlich nur die Andeutung des tieferen Gehaltes, der herauszuholen, des Schatzes, der zu heben ist*] (Bauer-Lechner, 32/27). Mahler's reference to the capability of music to express more than language is not original: Mendelssohn and many other nineteenth-century composers made this same claim for music. What is unique about Mahler's description of his process of Lied composition is the manner in which the text is a mere indication, or hint, of a deeper meaning – a meaning which may exist in more than one sign-system or code.

This approach to a poem is neither imitative nor systematic. In a response to Natalie Bauer-Lechner's inquiry as to how he set a text to music Mahler replied: "One minute it is the poem that is the inspiration, the next it is the melody, I often begin in the middle, often at the beginning, sometimes even at the end, and the rest gradually falls into place until it develops into a complete whole" [*Bald gibt das Gedicht den Anstoss, bald die Melodie. Oft fange ich in der Mitte, oft am Anfang, zuweilen auch am Ende an, und das übrige schliesst sich machher dran und drum herum, bis es sich sum Ganzen rundet und vollendet*] (33/29). Mahler claims it is the work of the composer to extract and develop this hidden significance of a poem: "It's a strange process! Without knowing at first where it is leading, you find yourself pushed further and further beyond the bounds of the original form, whose potentialities lay hidden within it, like the plant within the seed" [*Es ist ein seltsmer Vorgang! Ohne dass man anfangs weiss, wohin es führt, fühlt man sich immer weiter und weiter über die ursprüngliche Form hinaus getrieben, deren reicher Gehalt doch, wie die Pflanze im Samenkorn, unbewusst in ihr verborgen lag*] (Bauer-Lechner, 32/27).

Mahler's treatment of *Ich bin der Welt abhanden gekommen* as a piano-Lied is emblematic of the abductive nature of his comments about Lied composition. The entire setting is based upon the pitch-configuration C-D-F-A and various versions of this sonority stated in different inversions or at different pitch-levels. This pitch-configuration is outlined in the right-hand of the piano in measures 1–3 and stated vertically for the first time in measure 10. Depending upon the context of its use, this pitch-configuration may be understood as a D minor seventh chord in first inversion or as an F major triad in second inversion with an added sixth. In other words, this sonority may be heard as alluding to two keys at once, F major and/or D minor. Regardless of whether the reference is to the tonic major, its relative minor, or both, the sonority itself is an unstable one which begs for resolution by way of an appogiatura.

No doubt, the instability of this sonority reflects the ambivalent tone of the narrator toward the state of alienation examined in the poem. Likewise, the movement from dissonance to consonance which characterizes the appogiatura also characterizes the gradual resignation of the narrator toward this state of alienation. However, ambivalence and alienation are merely the poetic concepts which serve as indicators for Mahler's setting. Once Mahler makes the musical inference that this sonority and the voice-leading device which it implies are an adequate metaphor for the poem, he bases the development of the song solely on the musical properties of this inference.

If the Lied composition of the First and Second Berlin Schools represents a balance between poetic and musical means of expression, and the songs of Schubert and Schumann embody a hegemony of music in relation to poetry, then Gustav Mahler's songs represent an almost total indifference to the formal or semantic properties of the poem, and, instead, focus upon the musical development of the hidden implications of the text which they treat. Simply put, Reichardt and Zelter focus upon the poem *qua* poem, Schubert and Schumann focus upon the meaning of the poem, and Mahler focuses upon the polysemiotic implications which lay hidden within the poem and may be developed in the code of music.

MAHLER'S MUSICAL OVER-CODING OF RÜCKERT'S TEXT

Any time poetry and music are united in song, a conflation of codes occurs. This conflation results in an expression in which the characteristics of the code of music may amplify, de-emphasize, or ignore certain elements expressed in the verbal code of the poem. This conflation of codes and the hermeneutic function of music in relation to the text of the poem constitute an example of *over-coding* as described by Umberto Eco in his *Theory of Semiotics*. According to Eco, over-coding is a two-stage process in which: (1) a basic code establishes that a grammatical proposition is understandable, and (2) with the imposition of another code, a code which is laid over the prior proposition as it were, it becomes evident that the proposition in question must be used under certain circumstances and with certain connotations (133). When applied to the phenomenon of song, the two-stage process of over-coding takes place in the following manner: (1) the verbal code establishes that a poem exists and has some conventional meaning, and (2) the musico-poetic expression that is created by imposing the code of music on the poem must subsequently be understood as a musical expression able to convey certain musical connotations. This two-stage process of over-coding seems to adequately describe the manner in which any pre-existent text is set to music.

Eco then observes that the process of over-coding may culminate in one of two ways: it either analyzes the prior proposition or it adds to its meaning. He claims, "the process of over-coding may analyze the units of the original code into more analytical entities," or "the process may assign additional meanings to more macroscopic strings of expression" (134). When applied to the phenomenon of song, these observations further illuminate the differences between the inductive approach to Lieder over and against the abductive approach practiced by Mahler. In the Lieder of Schubert and Schumann, the musical setting seems to create a typology or typologies of musico-poetic signs which serve as commentary upon the meaning of the poetic text. Yet in Mahler's Lieder the setting – which is based upon and develops a musical inference or hypothesis – constitutes a parallel discourse which may at times interact with the poetic text, but which is also capable of signification in its own right.

Mahler over-codes Rückert's *Ich bin der Welt abhanden gekommen* three different times. The first instance of over-coding is the piano-

Lied composed on or about 16 August 1901 which has already been discussed. The second instance is an orchestral treatment of this same text in which Mahler transposes the piano-Lied from F major/D minor down to Eb major/C minor. This orchestral song is a literal rendering of the piano-Lied in which no part of the vocal line or accompaniment is modified. The instrumentation is for oboe, English horn, two clarinets, two bassoons, two horns, harp, and strings. The most remarkable aspect of this setting is Mahler's juxtaposition of musical timbres. In his handling of the first strophe of the poem the winds predominate (measures 1–27), while in the treatment of the second strophe (measures 28–39) the string texture is emphasized. The final strophe (measures 40–67) begins with triplet arpeggios in the harp and thirds in the clarinets, and it concludes with the English horn and the first violins in an appogiatura-laden dialogue. As was the case with the piano-Lied, this orchestral treatment is governed solely by musical considerations. There is nothing in the first strophe of text which demands a wind timbre nor is there anything in the second strophe which dictates a string timbre; in fact the three strophes of the poem display remarkably little narrative development. Rather, Mahler, the master orchestrator, is spinning out musical ideas and impulses in his orchestration of the piano-Lied. The third over-coding of Rückert's text is the Adagietto of Mahler's Symphony No. 5 in which the text of Rückert's poem is jettisoned and elements of both the piano-Lied and the orchestral song are developed in a purely instrumental context. In this over-coding the musical content of Mahler's previous treatments of *Ich bin der Welt abhanden gekommen* is combined with musical material from the second song of Mahler's *Kindertotenlieder, Nun seh' ich wohl, warum so dunkle Flammen*, another Rückert text examining the theme of death which is, itself, in the third phase of over-coding. This orchestral movement in the key of F major/D minor, scored for strings and harp, retains the structural significance of the C-D-F-A sonority and the appogiaturas which it implies. Likewise, it carries over the use of the harp for the arpeggiated triplets.

When viewed from the perspective of Charles Sanders Peirce's triadic model of signification, the three phases of Mahler's over-coding of Rückert's poem *Ich bin der Welt abhanden gekommen* may be diagrammed as follows:

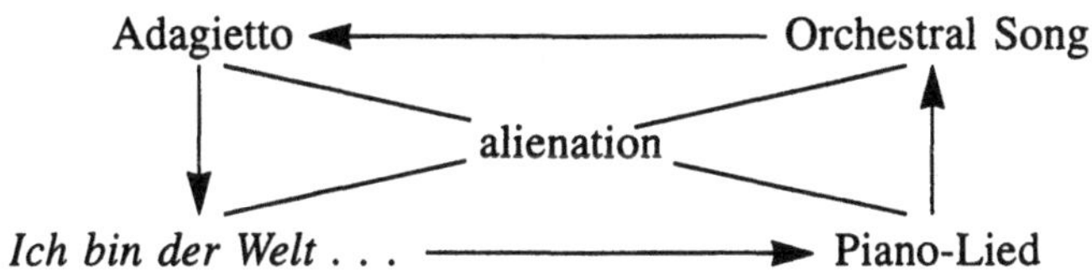

This model shows the manner in which Mahler's abductive approach to song composition can generate a purely instrumental composition capable of signification in its own right. The poem as originary representamen gives rise to a piano-Lied as interpretant both of which stand in the same relation to the object, alienation, which is the experience examined in the poem. This piano-Lied then functions as a representamen in a new triad in which its interpretant is the orchestral song, and again both the piano-Lied as representamen and the orchestral song as interpretant stand in the same relation to the object of the poem. Thirdly, the orchestral song takes on the role of representamen and the Adagietto of the Symphony No. 5 serves as interpretant, both of which stand in the same relation to the object of alienation. Finally, the Adagietto of Mahler's Symphony No. 5 serves as representamen and this time the Rückert poem which originated the entire process acts as an interpretant, both of them standing in the same relation to the object of alienation.

MAHLER'S MUSICAL CATACHRESIS

While the above diagram seems to adequately represent the process which generated these three different compositions by Mahler, there is something destabilizing in the fact that the most likely interpretant for the Adagietto of the Symphony No. 5 is the very same representamen which initiated this complex semiotic process. The result is that Mahler gives the listener no solid ground upon which to base an interpretation. Instead he or she must ask: Do I use the orchestral movement to understand the poem or do I use the poem to interpret the orchestral movement? i.e. does this movement function as a musical interpretant of Rückert's text or its the Rückert text a poetic interpretant of the symphonic movement?

Such a destabilizing question arises because the Adagietto is, itself, a mixed metaphor, or *catachresis*, in which a musical expression has been used as a substitute for poetic content. In his *De ratione dicendi* Cicero defines catachresis in the following manner: "Catachresis is the inexact use of a like and kindred word in place of the precise and proper one" [*simili et propinquo pro certo et proprio*] and he continues to assert that a catachresis has occurred when "words of kindred, but not identical meaning have been transferred on the principle of inexact use" [*ratione abusionis esse traducta*] (342–343). Mahler's musical catachresis, in which the purely instrumental Adagietto at once stands for the content of Rückert's *Ich bin der Welt abhanden gekommen* and, at the same time, possesses its own purely musical significance, seems to operate on this principle of *abusio*, or inexact use.

What, then, is the result of this complex compositional process carried out by Mahler and what claims about Mahler's music in general does the explication of this process allow? The answer to both of these questions is: new and valuable insights into the ironic nature of Mahler's music. As has been demonstrated, the process is one in which Mahler makes an abductive inference about a poetic text in terms of a musical metaphor; he then proceeds to over-code the text with this musical metaphor and its development, and concludes by allowing the musical metaphor to stand alone as an autonomous instrumental movement; however, this seemingly evolutionary process is undercut by the possibility that the very text which initiated the progression may be the most adequate interpretant of the result of the progression. Thus the generative process is reduced to a tautology.

The irony of this situation is not the kind of localized irony which is so obvious in Mahler's music, e.g. the Ländler rhythms, brass bands, military signals, and folk tunes, which he interpolates into the context of art-music. Rather, the irony which this process embodies is a more pervasive, and perhaps more pernicious, irony which challenges the ability of any expression to communicate in an efficacious manner. It is the same irony which lay behind Peirce's assertion that we know our world and even ourselves only by means of signs. It is this same irony which informed Saussure's metaphor of language as a chess game. Moreover, it is the same irony which motivated Wittgenstein to conclude his *Tractatus Logico-Philosophicus* with the claim, "what we cannot speak about we must pass over in silence" [*Wovon man nicht sprechen kann, darüber muss man schweigen*] (150–51). I, for one, am grateful that Mahler chose, not silence, but instead, such an eloquent and exquisitely ironic mode of musico-poetic discourse as that which is embodied in his various treatments of Rückert's *Ich bin der Welt abhanden gekommen.*

Goshen College

NOTES

[1] Peirce states: "Nothing has so much contributed to present chaotic or erroneous ideas of the logic of science as failure to distinguish the essentially different character of different elements of scientific reasoning; and one of the worst of these confusions, as well as one of the commonest, consists in regarding abduction and induction together (often mixed also with deduction) as a simple argument" (8.228).

[2] For an extended discussion of the *intersemiotic qualities* shared by both poem and musical setting in Schumann's Op. 39, see my *Gesture, Sign, and Song.*

REFERENCES

Bauer-Lechner, Natalie, *Erinnerungen an Gustav Mahler*, ed. H. Killian (Hamburg: Karl Dieter Wagner, 1984).

Cicero, *De ratione dicendi*, trans. H. Caplan (London: Heinemann, 1954).

Eco, Umberto, *A Theory of Semiotics* (Bloomington: Indiana UP, 1976).

Hecker, Max, *Der Briefwechsel zwischen Goethe und Zelter* (Leipzig: Insel-Verlag, 1913.

Kramer, Lawrence, *Music and Poetry: The Nineteenth Century and After* (Berkeley: U of California Press, 1984).

Mosley, David L., *Gesture, Sign, and Song* (New York: Peter Lang, 1990).

Peirce, Charles Sanders, *Collected Papers* I–IV, ed. Hartshorne and Weiss (Cambridge: Harvard UP, 1931–1958).

Saussure, Ferdinand, *Cours de linguistique générale* (Paris: Payot, 1916).

Schumann, Robert de, *Geseammelte Schriften Über Musik und Musiker* (Leipzig: Wigand, 1871).

Wittgenstein, Ludwig, *Tractatus Logico-Philosophicus* (London: Routledge, 1961).

SITANSU RAY

MADANA (THE MYTHICAL LOVE-GOD) IN KALIDASA AND TAGORE

The mythical allegory of Madana, the god of love, has been continually appearing in the realm of our creativity and culture since ancient times. The source of Madana can be traced back not in the *Vedas* but in the *Puranas* (the Hindu mythologies), especially in the *Kalika Purana* and also in the *Brahmavaivarta purana, Siva Purana* and *Matsya Purana*. The aim of this paper is to sketch some selected depictions of Madana as found in Kalidasa's and Rabindranath's works, with a view to reviewing these in our present context.

But before that, let us take into account the various other names of Madana and their specific etymological meanings so as to reveal their allegorical purports, semiotic significances and, furthermore, their mythological references and phenomenological grasps and reflections. The word *Madana* means embodiment and diffusion of amorousness. The name *Kamadeva* means god of love and lust. The names *Ananga* and *Atanu* both mean the incorporeal one, the God having no body of his own but being ever-present in males and females as the tremendous desire for uniting with each other. Almost the same meaning is inherent in the name *Manasija*, i.e., the god who is born in the core of the psyche. The name *Manmatha* means the God who is born of churning the mind, and who churns the mind as well. The name *Vasantasakha* means the friend of *Vasanta*, i.e., the spring season. The name *Ratipati* bears the further identity of Madana as the consort and husband of Rati, who is known as the goddess of copulation. The name *Makaravahana* signifies that Madana rides on *Makara*, the mythological aquatic animal, which is the astrological sign Capricorn. The name *Minaketu* means the God whose flag bears the symbol of fish. Another name of Madana is *Kusumadhanu* or *Pushpadhanu*, which means the God whose bow is made of flowers. He is also named *Panchavana* or *Panchashara*, i.e., owning five arrows, namely enchantment, excitement, absorption, heating and stupefaction. There are other versions of these five arrows, all denoting erotic symptoms of different degrees, such as *harshan* or joyousness, *rochan* or liking, *mohan* or infatuation, *shoshan* or suction and *Maaran* or erotic thrust. These are all amorous emotions or amorous

M. Kronegger and A-T. Tymieniecka (eds.), Analecta Husserliana XLII, pp. 303–308.

acts personified. Some other version provides names of five flowers as Madana's five arrows, such as *aravinda* or the lotus, *ashoka* or the tropical flamboyant, *chuta* or mango twigs with blossoms (shoots), *navamallika* or a kind of jasmine and *nilotpala* or the blue lotus. Last but not least, Madana is also called *Kandarpa*, which means the god who quelled the ascetical vanity of Lord Shiva, causing his union with Parvati, the daughter of Himalaya and Menaka.[1]

Kumara or Kartikeya, who became the Commander-in-Chief of the Army of Heaven, was the son of Shiva and Parvati. Kalidasa's (circa the fifth century A.D.) poetical work *Kumarasambhavam*[2] (the birth of Kumara), consisting of seventeen cantos, is based on Madana's role in the wedding of Shiva and Parvati and the birth of Kumara thereafter. The relevant portion of the story is as follows:

Once upon a time the Gods and Goddesses of the Heaven were very much disturbed by the wicked Tarakasura (the demon called Taraka). All of them resorted to Brahma, the Creator, for a solution. Brahma suggested that the son to be born of Shiva's semen would be able to defeat Tarakasura, and that the great ascetic Lord Shiva's semen could be fruitfully conceived only by his ardent devotee Parvati. Since Shiva was an ascetic, it seemed very difficult to nuptially unite Shiva and Parvati. So, Brahma entrusted Madana with this task.

Madana, accompanied by Rati and Vasanta, went to the hermitage of Shiva. All on a sudden, untimely spring appeared there. The trees, creepers, flowers, animals, birds and nymphs turned colourful and amorous. But Shiva was still unperturbed in his meditation. Parvati was worshipping him as usual. He blessed her with unattached affection as he had been doing every day. At that very moment, Madana applied all his tricks to pour passion into Shiva's mind. Shiva was seized with a passionate attraction toward Parvati's lips. Suddenly he checked himself, smelling some plot behind it. As soon as he saw Madana playing tricks he became furious. Fire came out of his third eye and Madana was turned into ashes. Rati lost her senses, Shiva left the place, and Himalaya took her daughter home, as she had been severely shocked.[3]

After regaining her senses, Rati burst into tears, mourned her dead husband, cried aloud and was ready to enter her husband's pyre. Vasanta consoled and protected her. At that moment, a celestial voice was heard from the sky, declaring that Madana would regain his body and life only after the destined marriage of Shiva and Parvati.[4]

Parvati resorted to hard austere worship. In the long run, Shiva was pleased and accepted her as his wife. In the midst of a blissful ceremony of music, dance and theatre of the divine artists, Shiva and Parvati got married. Madana returned to life and was united with Rati for good. Shiva, the eternal masculine and his consort Parvati, the eternal feminine, became engaged in all sorts of amorous enjoyment, pleasure and intercourse. Kalidasa's poetical depiction of them is unique. In the course of time Kartikeya was born, who was destined to be the Commander-in-Chief of the divine regiment for defeating the demon in order to restore peace. The classic grandeur of Kalidasa's *Kumarasambhavam* never seems to go out of style. Rather, the modern readers' allegorico-phenomenological approach to the episode reveals its pragmatic significance in human affairs of all ages. Even its hyperbolic exaggeration is but metaphorical thrashing out of human emotions and conditions.

Radindranath Tagore (1861–1941) was a voracious reader of all ancient, medieval and contemporary books, and his own works reflect these with newer and newer dimensions of his own poetic vision. "*Madana-bhasmer Purve*" (Before the Burning of Madana) and "*Madana-bhasmer Par*" (After the Episode of the Burning of Madana) are two excellent poems written by Tagore.[5] The theme of Madana returns anew to the reader. The effect of Madana is perpetual in mankind, in the animal kingdom and also in our natural surroundings. He cannot be burned. The burning episode is just a transitory phase signifying temporary suspension of erotic feeling. Madana's power is more diffused and more intense in both the sexes after that temporary suspension.

Panchashare dagdha kore korechho e ki Sannyasi, Visvamay diycchho tare chhadaye!
O Sannyasi, what have you done by burning Panchashara? You have rather surcharged the whole world with his spirit.[6]

The depiction of Madana in the short lyrical drama *Chitrangada*[7] (1892) is a bit different. Chitrangada, Arjun, Madana and Vasanta are the four characters in it. The source of the story is the *Mahabharata*. Chitrangada, the Princess of Manipur, was brought up like a prince with training in statesmaship and archery. Instead of caring for feminine charm, she developed a hero-like personality in her attire and movements. While hunting in a forest, she met the great hero Arjuna. Utter negligence and shocking rejection from him made Chitrangada repentant and restless. Actually Arjuna was following his vow of asceticism. Chitrangada turned to Madana and Vasanta, and prayed for at least temporary beauty and feminine charm so as to attract Arjuna. Her prayer was granted. She became extremely beautiful with all her feminine charms, but just for a year. She succeeded in distracting and enchanting Arjuna. Both of them became engaged in amorous enjoyment in full swing until Arjuna became fatigued and Chitrangada remorseful for distracting the hero with her false outer charm and sensuality. Once again she invoked Madana to take back all her alluring and deceitful charms. Madana granted her wish and adorned her with ideal womanhood. With a newer sense of love and companionship, Arjuna and Chitrangada were united.

The drama was thoroughly remodelled as *Nrityanatya Chitrangada*[8] (*The Dance-Drama Chitrangada*, 1936) by Tagore. By virtue of its music and dance, it is wonderfully successful on stage. The prehistorical mythology is renewed for us in the forms of literature and the performing arts. The allegorical significance of the mythology is obviously relevant in our modern days in the contexts of the feminist movement, woman's changing position in family and society, and the psycho-physical relationship between man and woman. Detailed interpretations may differ among readers and spectators since dynamic reciprocity between the sexes cannot be reduced to a set of norms of rules or ideals. Yet the concluding song of Chitrangada is very much important. The ultimate woman in Chitrangada wants from her consort Arjuna neither romantic eulogy nor lavish subjugation, but real companionship in the pains and pleasures of a mutual life. So, Madana's gifts are not limited to mere beauty and sensuality but extend towards blissful fulfilment which is to be achieved and shared by both the sexes. As Chitrangada unveils herself, Arjuna receives her favorably. After breaking the shell of allegory, we also discover and achieve the precious gem of Tagorean realization. Chitrangada and Arjuna may be taken as representations of the woman and man of all the ages of the past, present and future.

Bhairava or Maharudra (the other names of the Lord Shiva) destroyed Madana's embodiment because of his cunning enticement, shallow lasciviousness and gross sexuality. The revitalized Madana is identified with Bhairava himself, whose love does not require any secret enticement. Sheer lustfulness is destructive for mankind. That is why it must be done away with. Love adorned with goodness (Shiva) and bliss should be our ideal. This very idea is the purport of the grand historico-imaginative drama *Tapati*[9] (1929). Tapati means the girl who worships the sun god. In this drama, Tapati is Sumitra, the queen of Jalandhar, the king being Vikramadeva. Their union was never happy. Vikramadeva experienced excessive infatuation with Sumitra. He actually abducted her from the royal family of Kashmir under the pretext of winning her in a battle against Kashmir. Yet he never entrusted her with the dignity of royal chores. The undercurrent of revolt in Sumitra led her back to Kashmir, only to sacrifice her life to Martandadeva (the sun god).

The drama starts with the song *Sarva kharvatare dahe tava krodhadaha* (The Fire of Your Wrath Burns Down All Frivolity), sung in chorus as a preamble to the worship of Minaketu (Madana), arranged by Vikramadeva in the courtyard of the royal temple of Bhairava in Jalandhar. The worship of Minaketu symbolizes overindulgence in carnal desires.

The prudent lady Bipasha, the close intimate of Sumitra, brings forth melodious relief by virtue of her songs in the midst of unnerving tensions and tragic events. Some of the songs are addressed to Pushpadhanu (Madana), some to Nataraja (Shiva) and some relate to various scenes.

Shunning the amorous aspect and highlighting the juvenile and youthful joy of the spring season, Tagore introduced *Vasantotsava* (the spring festival) at Santiniketan. Each year it is observed here with music and dance on the full moon, day and night, during the spring.

Tagore's dramas *Phalguni*[10] (of the Bengali month Phalgun, February-March or early spring) and *Vasanta*[11] (The Spring season) are two excellent dramas. Since they were designed to be staged mainly by students and teachers, their amorous side has been eliminated. Yet the juvenile joy of communion with nature is at the core of these dramas.

Tagore's *Nataraja: Riturangashala*[12] is another kind of work consisting of numbers of recitations and songs representing the six seasons, the last one being spring.

All these dramas and works are allegorico-metaphorical in nature, personifying natural objects and events and thereby reasserting man's relationship with nature. Love is obviously present, but in a serene form.

Visva-Bharati University

NOTES

[1] See Bangiya Sahitya Parishad, *Bharatakosha*, Part V (Calcutta, 1973), pp. 277–8. *Bangla Visvakosha*, ed. N. Basu (New Delhi: B. R. Publishing Corporation, 1988), Part 14, pp. 24–26.

Also consult the Sanskrit *Amarakosha* and different dictionaries (Bengali to English).

All the names of Madana have come to Bengali from Sanskrit. Most are compound words revealing mythological allusions and allegories.

[2] There are various editions of *Kumarasambhavam* along with English translations of the verses.

[3] See the third canto of *Kumarasambhavam.*

[4] The details are there in the fourth canto of *Kumarasambhavam.*

[5] Rabindranath Tagore, *Kalpana* (Imagination) (1900), *Rabindra-Rachanavali* (collection of Tagore's works, henceforth to be written as *RR*) Vol. I, Birth-Centenary Edition (Govt. of West Bengal, 1961), pp. 701–3.

[6] R. Tagore, "*Madan-bhasmer Par*", *op. cit.*, p. 702.

[7] R. Tagore, *Chitrangada* (Calcutta: Visva-Bharati Publishing Department, 1966).

[8] R. Tagore, *Nrityanatya Chitrangada* (Calcutta: Visva-Bharati Publishing Department, edition 1358 B.S. 1951).

[9] *RR* 6, pp. 1031–1090.

[10] 1916; edition with musical notation 1355 B.S. (1948). Thereafter various editions have appeared from the Visva-Bharati Publishing Department.

[11] *RR* 6, pp. 441–488. 1329 B.S. (1923). Visva-Bharati editions started from 1337 B.S. onwards. *RR* 5, pp. 585–601.

[12] *RR* 5, pp. 619–663.

ANNEX

A. ZVIE BAR-ON

FREEDOM AND CREATIVITY

I would like to treat the concept of freedom by distinguishing its levels or grades and analysing the relations between them. It is also possible to speak of different positions about freedom and to arrange them in an order of logical succession. My argument, about these positions are an attempt to discover the phenomenology and dialectic of the concept of human freedom. A systematic analysis of the concept of freedom will lead me to an interpretation of the concept of creativity.

I shall distinguish between *four* basic positions about freedom:

Position (a): One is free if and only if he can fulfill his desires without any interference from outside.

Position (b): One is free if and only if
1. as in (a), and
2. the fulfillment of the desires is controlled by one's reason.

Position (c): One is free if and only if
1. as in (b), and
2. one adopts rules (principles, norms, theories) of action and behaviour and applies them in active life, as for example being moral or religious or adopting some social ideology.

Position (d): One is free if and only if
1. as in (c), and
2. one lives a creative life.

Now let us consider these four positions or grades of freedom and the relations between them.

Position (a) refers to the simplest, the most elementary kind of freedom. In evaluating the absence of external intervention it comes close to the attitude of the so called "soft determinist" (in accordance with William James's distinction between "soft" and "hard" determinism).[1]

In the literature this attitude is sometimes characterized as "negative freedom", in contradistinction to "positive freedom" which will be considered later. One could also speak here about the necessary, in contradistinction to the sufficient condition of freedom.

Now, there is a difference between position (a) and "negative freedom". In the definition of position (a) *desire* is mentioned as the

M. Kronegger and A-T. Tymieniecka (eds.), Analecta Husserliana XLII, pp. 311–318.

only factor of action while in the usual definition of negative freedom there is no such limitation. The conception of negative freedom assumes various kinds of motivation including "enlightened desire", that is a desire guided by reason, which belongs in our classification to position (b).

The point that desire is the only motive in our active life can be criticised. "Desire" is a polymorphous expression. In the changing contexts of the everyday speech various connotations are connected with it. Our way of using "desire" is not said to cover all these connotations. It is therefore of a limited use. Generally it is possible to characterise its use – as some psychologists and sociologist, recommend – by one or more of the following concepts: (1) organic needs; (2) instincts, impulses; (3) innate dispositions.[2]

Organic needs are reflected in the permanent, natural tendency, built-in our organism to perform activities necessary for its existence and development, like eating, sleeping, breathing, etc.

Curiosity, or the natural tendency to acquire knowledge, is often cited as an example of the *second* category of desire, together with the sexual drive, parental attachment to children, acquisition of material belongings, construction as well as destruction, self-confirmation, etc.

The *third* category of desire is harder to define and exemplify, than the previous two. Psychologists and sociologists differ widely between each other in this respect. The following example, however, may give us a fair general idea of the species of innate inclinations and dispositions which are not supposed to fall under the categories of organic needs, like food and sleep, or of instincts, like sex and curiosity: sympathy for one's fellowman; the tendency to participate in social or political life; the delight one finds in playing games; the inclination to form habits, to adopt beliefs, etc.

The freedom of position (a) is said to be assured by the condition that no outside factor would prevent the fulfillment of our desires. Certainly, if a desire is such that in its fulfillment another person is involved, as for example the sexual urge or parents' attachment to their children, this second part is not to be considered "an external factor".

However, a quick look at human nature will reveal that there are also *internal* factors which could prevent an undisturbed, harmonious gratification of a desire. To mention two of these factors will suffice: (1) An opposition or contrast between desires of the same person may arise; for example, the sexual drive may, in certain circumstances, come into conflict with the parental attachment to children; the tendency to self-assurance does not always coincide with the inclination to imitate the behaviour of others; the instinct to ruin things does not generally live in peace with the constructive instinct, etc.

(2) The exaggerated gratification of a desire, in particular with concern to food and the sex, may cause harm to the organism and even lead to its ruin. Also, the desires, together with the inclination to create habits, may lead to addiction, that is, to making a person a slave of his desires.

There is, therefore, a sense in which position (a) is unsatisfactory. To forgo this we shall proceed to position (b), the conception that freedom comes to fore when desire is guided by reason.

Unfortunately, "reason" is also a term the definition of which is extremely difficult. For the definition of reason Aristotle uses three concepts: theoretical wisdom, practical wisdom and technical skill. In Kant's writings we find a slightly different terminology. Here there is a triple distinction between theoretical reason (including "understanding"; Kant apparently introduced the systematic distinction between "understanding" [*Verstand*] and "reason" [*Vernunft*]), practical reason and skill (*Geschicklichkeit*).

It seems to me that Alfred North Whitehead gave the best key to our problem. He distinguished between the understanding of the fox and the reason of Plato. The position (b) needs the practical, foxy reason which would arbitrate between the desires conflicting with each other, curb those the gratification of which could endanger the existence or well-being of the organism, and will help the organism, the personality, to overpower the external factors of compulsion, hindrance and constraints.

It becomes clear that the freedom of position (b) is connected with concepts of self-restraint and rationality, as well as with the idea of internal and external factors of human existence. However, these concepts are now richer and more complex than what was assumed in connection with position (b). A more penetrating analysis of these concepts will lead us to the positions (c) and (d). I shall use Popper's theory of the Three Worlds in this analysis. First, a few words about that theory which presents our world as constituted by three sub-worlds, which Popper calls "World 1", "World 2", and "World 3."

World 1 is, as he says, "that of stones, threes and fields of physical forces".[3] They encompass all of Popper's three worlds. World 1 is mainly the world of physics, but it includes also the worlds of chemistry and biology. We have to be careful, however, not to confuse "chemistry" and "biology" as sciences (systems of theories, propositions expressing laws, descriptions of experiments and their analysis, etc.) with the *objects* which we usually call "material substances", on which experiments are performed and the explanation of which we expect from the theories and laws. Chemistry and biology *as sciences* do not belong to World 1.

World 2 is the psychological domain. It is being investigated by those who study the souls of human beings and of animals. It is the world of feelings and dispositions, of all experiences conscious and sub- or un-conscious. The term "psychological" may again be misleading. As the suffix "-logical" suggests, the more natural use of this term is as an adjective referring to laws, theories and scientific practices, rather than mental events and situations. But, again, we have to remember that laws and theories referring to, or resulting from mental events and processes *do not* belong to World 2, while those events and processes actually do; they constitute this world.

World 3 is all that the human mind creates. Popper says that although he includes in World 3 human societies and institutions, works of art and ethical values, etc., he limits himself in his considerations to the world of scientific libraries, of books, of scientific problems and of theories, including false ones. This limitation is irrelevant to our discussion. We retain the general definition of World 3: everything created by the human mind belongs to it.[4]

We have to be, however, careful here as well. When we think of a library we usually imagine a building with a collection of books and other equipment which can be used mainly for reading, studying. But not the building, nor the library equipment, nor books as bound sheets of paper covered with small items of paint, belong to World 3. Their proper place is World 1. Analogously, the mental events constituting the processes of reading and studying *do not* belong to World 3. Their place is World 2.

However, when we read a book, we are not interested mainly in the visual and other experiences of perceiving this physical object. We are interested in its content – say, in the scientific theory stated and explained in it. This content, expressed in the book by linguistic and other means, is an object of World 3. As such it has a specific ontic status, different from that of the objects of World 1 and World 2. It is subject to different regularities, limitations and evaluations. Moreover, it has specific relations with objects of the other two Worlds.

The objects of World 3 are products of culture, science and political organization, rather than of purely natural processes. They are *human products* and it is important to see how they are produced, how they can be described and explained. One of the main factors in the acts of production and creativity is *human language*. Popper writes on this point:

> Human language and human thought evolve together, in mutual interaction. Human language, admittedly, expresses human thought processes, that is, the World 2 objects. But it makes a very great difference to these subjective World 2 objects when they are formulated in an objective human language: there is a powerful feedback effect between human language and the human mind.
>
> This is mainly because a thought, once it is formulated in language, becomes an *object* outside ourselves. Such an object can then be inter-subjectively *criticized* – criticized by others as well as by ourselves. Intersubjective or objective criticism in this sense only emerges with human language; and with it emerges the human World 3, the world of objective standards and of the contents of our subjective thought processes.[5]

The comparison between human language and the language of other animals will reveal a still deeper aspect of Popper's theory. He writes:

> [T]his is not to belittle the gulf which separates the human brain from the animal brain, and human language from all other animal languages – from the dispositions which most higher animals possess to *express* their inner states and to *communicate* with other animals.
>
> Man has created human language, with its *descriptive* function and the value of truth, and with its *argumentative* function and the value of the validity of arguments, thus transcending animal languages with their merely expressive and communicative functions. With it man has created the objective World 3, something that there are only fairly remote analogues of in the animal kingdom. And with this he has produced a new world of civilization, of learning, of non-genetic growth: of growth that is not transmitted by the genetic code; of growth which depends not so much on natural selection as on selection based upon rational criticism.[6]

Now we are able to explain the deep meaning of human freedom, of the dimensions which tie up the concept of freedom with the passage from position (b) to positions (c) and (d).

First of all, we have now a much more complex and comprehensive concept of rationality than that which we worked with for the characterization of position (b). Now it is defined as critical thought or rational criticism. Such critical thought is said to be expressed by the adoption of general principles and their consistent application in active life. The rule represents a considerable range of concepts akin to each other, like norm, law, theory, principle, and in certain contexts also goal, ideal, value, etc.

We consider human action or human behaviour to be *rule-governed*, i.e., subject to rules, with the proviso that it is man who creates these rules, and also – at least sometimes – applies them sincerely and seriously in his active life.

Another important point is that for the freedom of an individual, in contradistinction to that of the whole of mankind, it is not necessary that he himself has created the rules. It is sufficient for him that these rules exist (in World 3) and that he is able to adopt and apply them. This fact gives the individual a tiny corner for critical thinking, which makes him free in the basic sense of the concept of freedom, free from the blind determination of forces independent of him.

To grasp this freedom rightly, we have to understand, first of all, the universal character of these rules and the rule-governed behaviour of man. A clear example of it is the situation of man as a political being in, say, the Aristotelian sense.

Man, as we know him, was born into a political system, a system of law, habits, institutions, etc., and as a rule he goes on living in such a system all his life. He may move, in certain circumstances, from one system into another, he may also try to change the system in which he lives, or that into which he has arrived. But as a human being he cannot – except perhaps in very particular circumstances, *imagined* rather than real – live in a politically empty space.

But clearly, the political system is not the only one which the human being belongs to. The linguistic activity of man, his use of a certain language with its specific functions – this too is an excellent example of rule-governed behaviour. So is the activity of a teacher, a scientist or any specialist whatsoever.

Prima facie, it is perhaps difficult to see clearly the aesthetic and art experience as rule-governed behaviour. The creative imagination – some may say – does not obey rules. However, the very line dividing art from non-art, in any of its domains, be it literature, the world of sounds or visual images – this line is clearly determined by certain criteria of evaluation, i.e., by rules which one has to internalise, sometimes in a long process of working as an apprentice to a recognized artist, or in other cases being systematically subjected to the influence of recognized works of art. To recognise a work of art as such is, basically, also a matter of fulfilling certain criteria, i.e., certain rules of evaluation.

We may assume that human life, as far as it can be so called, i.e., social and cultural activity, flows uninterruptedly through a considerable range of normative systems. Such a life has two dimensions of freedom: (1) the adoption of rules of evaluation which were established by others, and their application in the ramified experience of the person, and (2) which may be formulated in two sub-dimensions: (2′) a critical re-evaluation of existing rule-systems and their eventual improvement; (2″) abandonment of a system of existing rules and establishment of a new system in its place.

It seems to be that this last mode of rule-governed activity is very close to what could be called "human creativity". I believe this is the highest rung of freedom, especially when a whole system of norms and criteria, or at lėast a great part of such a system, is concerned. For example, there is the discovery of a new scientific theory, but such that after it gains corroboration and approval a new scientific discipline is established or an existing one is basically changed; there is the introduction of a new current or style in the arts like music or painting; there is establishment of a new political movement which, at the end of the process, will create a revolutionary change in the existing regime or establish a new regime; and there is a technological innovation which will cause a serious change in human life, etc.

These are examples in which creativity exists in its pure, perfect and most impressive form. However, creativity in its more modest forms may be found in all domains of human experience. Basically, it shows in any deviation from approved norm, on the assumption that it was accompanied by critical thought and constructive intention. When this happens, man rises to the highest rung of free action. The continuity of existence of the Popperian World 3 depends, partly, upon the appearance of a very great number of such deviations, which happen every day and every hour in all domains of human activity.

On the other hand, the continuity of existence of World 3 depends also on an opposite phenomenon: the *maintenance* of certain rules of the game, apparently more basic ones than those that can be changed by the creative drive. This would suggest the idea of the need for an equilibrium between the positions (c) and (d). This shows in what sense the positions (c) and (d) are two levels of freedom the second of which assimilated the first one.

It seems also that the above considerating exposes one more dimension of critical thought in alluding to the relation between the concept of freedom and that of responsibility. But this is a subject for a different discussion.

The Hebrew University of Jerusalem

NOTES

[1] See W. James, "The Dilemma of Determinism" in *The Writings of William James: A Comprehensive Edition* (The Modern Library, New York: 1968), pp. 587–610

[2] See for the example W. McDougal, *An Introduction to Social Psychology* (London: Methuen, 1963), Chs. 2 and 3.

[3] K.R. Popper, "Epistemology Without a Knowing Subject" and "On the Theory of the Objective Mind," in: *Objective Knowledge: An Evolutionary Approach* (Oxford: Clarendon Press, 1972), pp. 106–190

[4] K. R. Popper, "Indeterminism is Not Enough: An Afterword," in *The Open Universe: An Argument for Indeterminism* (New Jersey: Rowman and Littlefield, 1982), pp. 113–130.

[5] *The Open Universe*, p. 118.

[6] *Ibid.*, p. 122.

INDEX OF NAMES

– Z –

Analecta Husserliana

The Yearbook of Phenomenological Research

Editor-in-Chief

Anna-Teresa Tymieniecka

The World Institute for Advanced Phenomenological Research and Learning, Belmont, Massachusetts, U.S.A.

1. Tymieniecka, A-T. (ed.), *Volume 1 of Analecta Husserliana.* 1971 ISBN 90-277-0171-7
2. Tymieniecka, A-T. (ed.), *The Later Husserl and the Idea of Phenomenology.* Idealism – Realism, Historicity and Nature. 1972 ISBN 90-277-0223-3
3. Tymieniecka, A-T. (ed.), *The Phenomenological Realism of the Possible Worlds.* The 'A Priori', Activity and Passivity of Consciousness, Phenomenology and Nature. 1974 ISBN 90-277-0426-0
4. Tymieniecka, A-T. (ed.), *Ingardeniana.* A Spectrum of Specialised Studies Establishing the Field of Research. 1976 ISBN 90-277-0628-X
5. Tymieniecka, A-T. (ed.), *The Crisis of Culture.* Steps to Reopen the Phenomenological Investigation of Man. 1976 ISBN 90-277-0632-8
6. Tymieniecka, A-T. (ed.), *The Self and the Other.* The Irreducible Element in Man, Part I. 1977 ISBN 90-277-0759-6
7. Tymieniecka, A-T. (ed.), *The Human Being in Action.* The Irreducible Element in Man, Part II. 1978 ISBN 90-277-0884-3
8. Nitta, Y. and Hirotaka Tatematsu (eds.), *Japanese Phenomenology.* Phenomenology as the Trans-cultural Philosophical Approach. 1979 ISBN 90-277-0924-6
9. Tymieniecka, A-T. (ed.), *The Teleologies in Husserlian Phenomenology.* The Irreducible Element in Man, Part III. 1979 ISBN 90-277-0981-5
10. Wojtyła, K., *The Acting Person.* Translated from Polish by A. Potocki. 1979 ISBN Hb 90-277-0969-6; Pb 90-277-0985-8
11. Ales Bello, A. (ed.), *The Great Chain of Being* and *Italian Phenomenology.* 1981 ISBN 90-277-1071-6
12. Tymieniecka, A-T. (ed.), *The Philosophical Reflection of Man in Literature.* Selected Papers from Several Conferences held by the International Society for Phenomenology and Literature in Cambridge, Massachusetts. Includes the essay by A-T. Tymieniecka, *Poetica Nova.* 1982 ISBN 90-277-1312-X
13. Kaelin, E. F., *The Unhappy Consciousness.* The Poetic Plight of Samuel Beckett. An Inquiry at the Intersection of Phenomenology and literature. 1981 ISBN 90-277-1313-8
14. Tymieniecka, A-T. (ed.), *The Phenomenology of Man and of the Human Condition.* Individualisation of Nature and the Human Being. (Part I:) Plotting

the Territory for Interdisciplinary Communication. 1983
Part II see below under Volume 21. ISBN 90-277-1447-9

15. Tymieniecka, A-T. and Calvin O. Schrag (eds.), *Foundations of Morality, Human Rights, and the Human Sciences.* Phenomenology in a Foundational Dialogue with Human Sciences. 1983 ISBN 90-277-1453-3
16. Tymieniecka, A-T. (ed.), *Soul and Body in Husserlian Phenomenology.* Man and Nature. 1983 ISBN 90-277-1518-1
17. Tymieniecka, A-T. (ed.), *Phenomenology of Life in a Dialogue Between Chinese and Occidental Philosophy.* 1984 ISBN 90-277-1620-X
18. Tymieniecka, A-T. (ed.), *The Existential Coordinates of the Human Condition: Poetic – Epic – Tragic.* The Literary Genre. 1984 ISBN 90-277-1702-8
19. Tymieniecka, A-T. (ed.), *Poetics of the Elements in the Human Condition.* (Part 1:) The Sea. From Elemental Stirrings to Symbolic Inspiration, Language, and Life-Significance in Literary Interpretation and Theory. 1985
For Part 2 and 3 *see below* under Volumes 23 and 28. ISBN 90-277-1906-3
20. Tymieniecka, A-T. (ed.), *The Moral Sense in the Communal Significance of Life.* Investigations in Phenomenological Praxeology: Psychiatric Therapeutics, Medical Ethics and Social Praxis within the Life- and Communal World. 1986
ISBN 90-277-2085-1
21. Tymieniecka, A-T. (ed.), *The Phenomenology of Man and of the Human Condition.* Part II: The Meeting Point Between Occidental and Oriental Philosophies. 1986 ISBN 90-277-2185-8
22. Tymieniecka, A-T. (ed.), *Morality within the Life- and Social World.* Interdisciplinary Phenomenology of the Authentic Life in the 'Moral Sense'. 1987
Sequel to Volumes 15 and 20. ISBN 90-277-2411-3
23. Tymieniecka, A-T. (ed.), *Poetics of the Elements in the Human Condition.* Part 2: The Airy Elements in Poetic Imagination. Breath, Breeze, Wind, Tempest, Thunder, Snow, Flame, Fire, Volcano… 1988 ISBN 90-277-2569-1
24. Tymieniecka, A-T., *Logos and Life.* Book I: Creative Experience and the Critique of Reason. 1988 ISBN Hb 90-277-2539-X; Pb 90-277-2540-3
25. Tymieniecka, A-T., *Logos and Life.* Book II: The Three Movements of the Soul. 1988 ISBN Hb 90-277-2556-X; Pb 90-277-2557-8
26. Kaelin, E. F. and Calvin O. Schrag (eds.), *American Phenomenology.* Origins and Developments. 1989 ISBN 90-277-2690-6
27. Tymieniecka, A-T. (ed.), *Man within his Life-World.* Contributions to Phenomenology by Scholars from East-Central Europe. 1989
ISBN 90-277-2767-8
28. Tymieniecka, A-T. (ed.), *The Elemental Passions of the Soul.* Poetics of the Elements in the Human Condition, Part 3. 1990 ISBN 0-7923-0180-3

Analecta Husserliana

29. Tymieniecka, A-T. (ed.), *Man's Self-Interpretation-in-Existence.* Phenomenology and Philosophy of Life. – Introducing the Spanish Perspective. 1990 ISBN 0-7923-0324-5
30. Rudnick, H. H. (ed.), *Ingardeniana II.* New Studies in the Philosophy of Roman Ingarden. With a New International Ingarden Bibliography. 1990 ISBN 0-7923-0627-9
31. Tymieniecka, A-T. (ed.), *The Moral Sense and Its Foundational Significance: Self, Person, Historicity, Community.* Phenomenological Praxeology and Psychiatry. 1990 ISBN 0-7923-0678-3
32. Kronegger, M. (ed.), *Phenomenology and Aesthetics.* Approaches to Comparative Literature and Other Arts. Homages to A-T. Tymieniecka. 1991 ISBN 0-7923-0738-0
33. Tymieniecka, A-T. (ed.), *Ingardeniana III.* Roman Ingarden's Aesthetics in a New Key and the Independent Approaches of Others: The Performing Arts, the Fine Arts, and Literature. 1991
Sequel to Volumes 4 and 30 ISBN 0-7923-1014-4
34. Tymieniecka, A-T. (ed.), *The Turning Points of the New Phenomenological Era.* Husserl Research – Drawing upon the Full Extent of His Development. 1991 ISBN 0-7923-1134-5
35. Tymieniecka, A-T. (ed.), *Husserlian Phenomenology in a New Key.* Intersubjectivity, Ethos, the Societal Sphere, Human Encounter, Pathos. 1991 ISBN 0-7923-1146-9
36. Tymieniecka, A-T. (ed.), *Husserl's Legacy in Phenomenological Philosophies.* New Approaches to Reason, Language, Hermeneutics, the Human Condition. 1991 ISBN 0-7923-1178-7
37. Tymieniecka, A-T. (ed.), *New Queries in Aesthetics and Metaphysics.* Time, Historicity, Art, Culture, Metaphysics, the Transnatural. 1991 ISBN 0-7923-1195-7
38. Tymieniecka, A-T. (ed.), *The Elemental Dialectic of Light and Darkness.* The Passions of the Soul in the Onto-Poiesis of Life. 1992 ISBN 0-7923-1601-0
39. Tymieniecka, A-T. (ed.), *Reason, Life, Culture, Part I.* Phenomenology in the Baltics. 1993 ISBN 0-7923-1902-8
40. Tymieniecka, A-T. (ed.), *Manifestations of Reason: Life, Historicity, Culture.* Reason, Life, Culture, Part II. Phenomenology in the Adriatic Countries. 1993 ISBN 0-7923-2215-0
41. Tymieniecka, A-T. (ed.), *Allegory Revisited.* Ideals of Mankind. 1994 ISBN 0-7923-2312-2

Analecta Husserliana

42. Kronegger, M. and Tymieniecka, A-T. (eds.), *Allegory Old and New.* In Literature, the Fine Arts, Music and Theatre, and Its Continuity in Culture. 1994 ISBN 0-7923-2348-3
43. Tymieniecka, A-T. (ed.): *From the Sacred to the Divine.* A New Phenomenological Approach. 1994 (forthcoming) ISBN 0-7923-2690-3

Kluwer Academic Publishers – Dordrecht / Boston / London

MIX
Papier aus verantwortungsvollen Quellen
Paper from responsible sources
FSC® C105338

If you have any concerns about our products,
you can contact us on
ProductSafety@springernature.com

In case Publisher is established outside the EU,
the EU authorized representative is:
Springer Nature Customer Service Center GmbH
Europaplatz 3, 69115 Heidelberg, Germany

Printed by Libri Plureos GmbH
in Hamburg, Germany